Mysterious Ways

Mysterious Ways

Contingency, Emergence, and Selection in Nature

JONATHAN D. PHILLIPS

University of Kentucky, Lexington, Kentucky,

United States of America

OXFORD
UNIVERSITY PRESS

Oxford University Press is a department of the University of Oxford.
It furthers the University's objective of excellence in research, scholarship,
and education by publishing worldwide. Oxford is a registered trade mark of
Oxford University Press in the UK and in certain other countries.

Published in the United States of America by Oxford University Press
198 Madison Avenue, New York, NY 10016, United States of America.

© Oxford University Press 2025

CIP data is on file at the Library of Congress

ISBN 9780197755099

DOI: 10.1093/9780197755129.001.0001

Printed by Marquis, Canada

MIX
Paper | Supporting
responsible forestry
FSC
www.fsc.org   FSC® C103567

CONTENTS

DEDICATION

In 2021, I published a book called *Landscape Evolution: Landforms, Ecosystems, Soils* with the scientific publisher Elsevier. It was meant to be a capstone, summarizing and synthesizing the main points of my research (and teaching) over the years—the coevolution of landforms, soils, hydrological systems, and ecosystems. It stressed—as does the book you are reading—the critical importance of place and history (along with general laws) in understanding and interpreting landscapes and the key roles of emergence and selection (including abiotic selection).

Landscape Evolution was written for a scientific audience—I tried to make it accessible to a range of scientists but not necessarily to the general public (civilians, as I refer to members of the public who are not Earth, environmental, or ecological scientists). The book is not perfect, and there were things I could have and maybe should have done differently, but I am satisfied with it (well, with everything other than its paltry sales). But afterward, three factors came into play.

First, many folks (well, many by my swamp-hermit standards: family, friends, bartenders, etc.) expressed a keen interest in what I had written about. But I had to warn them that the book would be very tough going for civilians—this is not at all a this-is-over-your-head because-I'm-smarter-than-you stance. Many of those folks were reading or writing things that I could not handle (e.g., computer systems engineering, urban planning, literary criticism, electro-cardio-physiology—not to mention blueprints, repair manuals, and wiring diagrams—all of which would be lost on me). So, I thought, maybe there's a civilian market for this stuff.

Second, I became possessed by a desire to get my ideas to a larger, broader audience and convinced of the importance of the subjects I discuss for the world at large.

And third, I realized that I am not done. While *Landscape Evolution* is an exclamation point on my career as an academic, and I retired from my increasingly tiresome (due mainly to an ever-burgeoning population of university administrators and to knuckle-head trustees and state legislators) university job, I could not stop doing research. I have an addiction, I guess.

So, *Mysterious Ways* is in part an effort to make the ideas from *Landscape Evolution* more accessible to a broader audience—not dumbing down (at least not to a level any dumber than they were before) but instead using plainer language and either "bumper-stickering" some of the technical details or moving them to endnotes or sidebars so as not to clutter up the main text. *Mysterious Ways* also incorporates some of the new research I've done since 2020 and some new takes on the earlier material. Finally, recognizing that effective ideas need to be tied up in a broader story, I've tried to do that with the mysterious ways theme. This is the idea that definite patterns in nature recur frequently in time and space but are not dictated or predicted by any general laws of physics, chemistry, biology, mathematics, systems theory, geography, geology, or anything else. These recurring patterns also apply to entities that have no intentionality or goals: rocks, dirt, water, air, carbon or nitrogen atoms, and even many biotic entities don't care; they don't want or try to achieve anything.

And yet they do. Why? How? These are the mysterious ways that can be explained by what I cover in this book and which turn out to be less mysterious than we thought.

My many, many intellectual debts are evident in the people mentioned in the text and cited in the endnotes. They also extend to my many teachers, mentors, collaborators, students, field assistants, and those I had the opportunity to assist. I appreciate Oxford University Press being willing to take this project on after several other publishers passed on it ("very interesting, but not consistent with our current publication plans" was a common response).

Since early in my career, my curiosity has been constantly rejuvenated by my now-grown children, Nate and Damien Phillips. Nate, in particular, was an eager and enthusiastic field companion and assistant, with the wasp stings, tick bites, scratches, and scrapes to prove it. My grandchildren, Caroline and Andy Phillips (born in 2014 and 2017), continue to inspire me, both in general and on our "Deeda adventures," as they've come to be called.

Damien was born as Bay Rochelle Phillips and designated female at birth. He came out to us as transgender when he was nineteen years old. His courage and his stubborn resistance to conformity and the bravery of many in the LGBTQ community who insist on being who they are, publicly, continue to be an inspiration. I have always been a bit out of the mainstream in my research ideas, but Damien's example has boldened me to be even more unconventional—not in terms of being controversial for the sake of being controversial, but being who I am, who I need to be as a researcher and writer.

When I retired from the university, my wife, Lynn Roche Phillips, had high hopes that I would spend some of my freed-up time on maintenance and improvements on our house and land. Why she thought I would suddenly embrace things I had avoided for years is beyond me, but hope springs eternal. Instead, I kept doing research and writing, and bless her, she has been nothing but encouraging and supportive.

More to the point, Lynn is the strongest person I've ever known. Damien/Bay was born three and a half months premature and weighed just 753 grams. My wife's mental, emotional, and physical strength in holding off premature labor for twelve critical days and dealing with ninety-nine days in neonatal intensive care is but one example. She has held the fort at home while I went off to do the fieldwork that much of this book is directly or indirectly based on. She has lifted me up and pushed me forward when that's what I needed and stood back out of my way when that was called for. She accomplished all of this all while pursuing her own remarkable career as the best university teacher in the whole damn world (yes, really). She is also an outstanding scholar, mentor, planning and environmental professional, and first-rate practitioner of community engagement—the kind that universities preach a lot about, but only a few, like Lynn, get their hands dirty doing.

Her love is the greatest gift I have ever been given.

I dedicate this book to Dr. Lynn Roche Phillips. I doubt there is anyone like her, but if there is, I dedicate it to them too.

Jonathan Phillips
New Bern, North Carolina
March 2024

Mysterious Ways

How does nature work?

That is the fundamental overarching question of science. Though an occasional foray into "why" is not out of bounds, the whys are primarily the province of religion and philosophy.

There are at least three dimensions to the "how" questions, one of which concerns stories. Based on observations and descriptions of *what* happens, can we produce (scientific) stories of *how* things happen? For instance, numerous observations of the formation of tropical cyclones (hurricanes, typhoons) lead to a story—an account of how such storms form. A second dimension concerns the fundamental underlying processes, mechanics, or dynamics of what happens. Using the same cyclone example, this includes the physics and chemistry of atmospheric pressure, air movement, heat fluxes, and state changes of water.

The third dimension of how nature works is the focus of this book. Given our stories about what happens and our knowledge of the underlying chemistry, physics, biology, etc. of the workings of individual natural systems, how does nature work the way it does in a broader sense, across geography, time, scientific disciplines, and natural phenomena? Underlying laws (by definition) work the same way everywhere and always. Why, then, do they produce so many different outcomes and pathways? On the one hand, all rivers, for instance, are governed by the same set of invariant laws governing how runoff is produced and translated, the

Mysterious Ways. Jonathan D. Phillips, Oxford University Press. © Oxford University Press (2025).
DOI: 10.1093/9780197755129.003.0001

forces exerted by flowing water, and so on. Why, then, even if you only look at rivers flowing through or across the same substrate (limestone, sand, etc.) do rivers take such wildly different forms and evolutionary pathways?

But on the other hand, given the different geological and climatic settings of rivers, the various biological influences, and other environmental factor variations, why do certain aspects of rivers (e.g., the topological and geometric properties of channels and networks or the recurrence of typical fluvial landforms) arise in vastly different environmental settings? A common answer throughout human existence is that this is how the creator (often said to work in mysterious ways) made it to be. This explanation is unsatisfying for those who don't believe in or are uncertain about the existence of a supreme being. And even for those whose who do believe (a category that includes a great many scientists), that answer is incomplete. Unless you stipulate a creator who is constantly at the controls, dictating what happens at levels from the subatomic to the planetary and beyond, there remains the question of *how* that creator makes it all happen. And really, even if there's someone or something always at the helm, what tools or controls does a constant controller use with respect to how they cause things to be manifested at Earth's surface? So, be you atheist or devoted believer or anything in between, the question lingers: how is the paradox of unlimited variety and recurring patterns and structures created and maintained?

This book focuses on how Earth surface systems—landscapes—arise, evolve, and function and how that relates to the "mysterious ways" of the world at large.

For a quick example of the sort of thing I'm concerned with here, consider a recent article by three prominent hydrological scientists—Hongkai Gao of East China Normal University, Fabricio Fenicia of the Swiss Federal Institute of Aquatic Science and Technology, and Hubert H. G. Savenije of Delft University of Technology in the Netherlands. They propose a radical reframing of approaches to watershed scale hydrology based on the idea that terrestrial ecosystems manipulate soil to manage water resources. Though I frame things a bit differently, I generally agree, as discussed later in Chapter 7. Without getting into the hydrological and ecological nuts and bolts, consider several quotes from the article: "Soils are embedded in the terrestrial ecosystems, which, through evolution and natural selection, have found ways to make best use of their resources." Later, "a terrestrial ecosystem adjusts its internal behavior to satisfy its needs, and it manipulates the substrate on which it grows." There are other examples, but you get the idea: in their view, ecosystems adapt and manage—and do so intentionally.[1]

Ecosystems include, by definition, communities of interacting organisms whose needs may coincide or overlap or which may be directly or indirectly at odds. While the individual organisms may perhaps be said to have intentionality, the community does not. Further, ecosystems include, by definition, abiotic aspects of the environment that can have no intentionality: water, rocks, and soil particles cannot care or know where they go, what they do, and whether they are supporting plants or other biota. At a finer level of detail, a carbon or oxygen or nitrogen atom is oblivious as to which molecules or compounds it belongs to. So how can an entity such as an ecosystem or a hydrological system be said to adapt or manage? How can its effects be intentional to any degree?

Mind you, what Gao, Fenicia, and Savenije claim about ecosystem and hydrological system evolution is correct with respect to what happens and why it works. But the "how" is the question—the source of mystery and the sort of mysterious way(s) I am concerned with in this book.

I have never met these authors, but I am confident that they do not believe and would not claim that communities, ecosystems, water, or soil have any conscious intentionality. But it is far simpler and clearer sometimes to speak of what happens in terms of some sort of broad goals (more on this later, too, especially in Chapter 5). Some cultures and languages attribute aliveness to the non-human and non-biological world. But in English, my one and only language, and in science, to attribute human traits such as goals and intentionality to other entities is considered anthropomorphism—a scientific and logical sin. Really, though, there's nothing wrong with repurposing language to help make sense of nature if it helps us understand. I did it quite often in teaching, using the scare-quote hand signals to indicate that a river or a soil does *literally* want or try to do anything.

So, I am not critiquing the use of language involving goals, purposeful adaptation, or intentionality as applied to environmental systems. What I am trying to do is reveal the mysterious ways in which nature makes it *appear* that such things are going on.

MYSTERIOUS WAYS

Most people have heard some version of the idiom "God works in mysterious ways." It is often evoked in a religious context, but the Lord working in mysterious ways is also deployed in a general and secular way to express the idea that the world, the universe, or nature functions in ways that we do not understand to produce outcomes we could not foresee.

Because the sentiment is typically expressed in terms of God's or the Lord's mysterious ways, I once assumed that the phrase originated in the Christian Bible. Not so. While the basic idea appears in the Bible and other religious texts (for example, Romans 11:33: "Oh, the depth of the riches and wisdom and knowledge of God! How unsearchable are his judgments and how inscrutable his ways!"), the *mysterious ways* phrase is linked to a hymn. William Cowper in 1773 composed a hymn called "Light Shining out of Darkness" about the hidden workings of the world, beginning with the line *God moves in a mysterious way*.

But it is not hard to find secular invocations of mysterious ways. Albert Einstein, who was often quite critical of religion, wrote in his book *Living Philosophies* that "the most beautiful thing we can experience is the mysterious. It is the source of all true art and all science. He to whom this emotion is a stranger, who can no longer pause to wonder and stand rapt in awe, is as good as dead: his eyes are closed." Henry David Thoreau, a patron saint of environmentalists, expressed in *Walden*, "We need the tonic of wildness. . . . At the same time that we are earnest to explore and learn all things, we require that all things be mysterious and unexplorable, that land and sea be indefinitely wild, unsurveyed and unfathomed by us because it is unfathomable." Outside the realms of science and religion, the song "Mysterious Ways" appeared on U2's album *Achtung Baby*.

Presumptuous though I may be, I do not presume to know, literally or metaphorically, the mind of nature, the universe, or any creator. Though I have certainly done my share of investigating the underlying processes of nature, you cannot expect any advances along those lines in this book. I have my own ideas on the "whys" of things on a philosophical and metaphysical level (quite complicated, confused, and informed by a hodgepodge of sources and difficult for me to articulate), but I will not get into that here.

This book *will* address the scientific whys, in the sense of why certain forms, structures, and patterns are effective, efficient, or advantageous. But that is only part of the story. Though we are learning more about ways that living things communicate with each other, ecosystems have no consciousness and cannot, for example, reach a decision to maximize entropy production or increase biodiversity. Rivers cannot care how, or even whether, they dissipate energy or transport sediment. Sand grains and wind are indifferent to the morphology of dunes, and soils cannot decide that they should develop dual-porosity hydrological regimes or vertical texture contrast profiles. There exist no laws that require any particular way to handle energy flows, biodiversity, energy dissipation, sediment transport, landform morphology, hydrological flows, or soil structure, but similar ways do, repeatedly, occur. Landscapes cannot purposefully cause these things to happen, yet they do, time and time again.

How?

Nature works in *mysterious ways*.

Here I will try to make them less mysterious. But first, some
   background.

WHAT IS A LANDSCAPE?

In my neck of the woods, the folks who manicure your lawn, trim
your shrubs, get rid of your fallen leaves, and such call themselves
landscapers. People also associate landscape with landscape architecture and engineering, site preparation, and the planning and
execution of gardens, parks, and grounds. This book is *not* about
landscape in any of those contexts. Rather, we are talking about
landscapes as the term is used in geomorphology, ecology, and
pedology (soil science).

In the study of Earth surface processes and landforms—
geomorphology—the basic unit of study is the landform, which can
be as small as an individual sand ripple on a dune or weathering cavity on a rock or as large as a mountain range or coastal
plain. A set of interconnected, geographically contiguous landforms
comprises a (geomorphic) landscape.

In the study of the interrelationships between organisms and
their environment—ecology—sets of interacting populations are
ecological communities, and communities along with their abiotic environment comprise ecosystems. A set of interconnected,
geographically contiguous ecosystems comprises an ecological landscape. As with landforms and geomorphic landscapes,
ecosystems and ecological landscapes can be demarcated at a
range of spatial scales. As these overlap, an ecosystem at one
scale may correspond to a landscape at a more detailed scale
(more on scale in Chapter 3).

In the study of soils as natural bodies (as opposed to as
a medium for plant growth or some other factor)—pedology—a
key unit of analysis is a pedon, which you can think of as a

three-dimensional version of a soil profile, from the ground surface to the underlying material from which the soil was formed (parent material). Interrelated, contiguous, more-or-less similar pedons make up a polypedon, which roughly corresponds in many cases with the soil types or map units shown on soil maps. Contiguous, interrelated polypedons comprise a soil landscape. While pedons and polypedons are more constrained in their spatial extent by the conventions of pedology and soil mapping, soil landscapes may also be defined at various scales, from a few polypedons or soil types to much larger aggregations.

For laypersons, a landscape (as opposed to landscaping) is essentially all the visible features of an area of countryside or land, often considered from an aesthetic perspective. The lay definition of a landscape is not at odds with the way I use the term here, but it is incomplete for our purposes, because we will consider the belowground and otherwise invisible parts, and we will not be directly concerned with aesthetics (though that is what originally inspired and continues to motivate many landscape scientists). Here we stick to the geomorphological, ecological, and pedological concept of landscapes and use landscape more or less synonymously with Earth surface systems (ESS), a general term encompassing ecological, geomorphological, hydrological, and soil systems, which (as we discuss below) are not just overlapping but also intricately, inextricably, and exquisitely interconnected.

In human geography and other social science and humanities disciplines, the definition of landscape has been vigorously debated at great length and in terms irrelevant to my purpose here. Please, let's not do that.

How We Understand Landscapes

A good way to start trying to understand something complex or complicated is to take it apart, as a way of breaking it into simpler, less-complicated pieces. Taking it apart is often figurative, of course, such as when we study or test or analyze different aspects of something one at a time, even if we do not literally disassemble or dissect. This approach—reductionism—is also useful in later stages of investigation to help work out the details.

Reductionism is the way science traditionally approached understanding Earth and its landscapes. Landscapes and ESS include the interacting systems at or near the literal solid surface of the planet, such as the topographic landscape, underlying geology, overlying atmosphere, living things (the biosphere), soils, and the various flows and storages of water, energy, nutrients, and such. With a massive and clever variety of reductionist techniques, we examine pieces of it all in the field and in the laboratory and break it down to its constituent elements as best we can. Anything is simpler when considered on a *ceteris parabus* (all other things being equal) basis. In the study of landscapes and ESS, I think of this reductionist science as constructing or identifying the pieces of a puzzle (actually a bunch of puzzles).

So, we begin to put together puzzles by studying ecosystems, hydrologic systems, soil systems, landform systems, climate systems, and so on. Climate, landforms, topography, soils, and ecosystems affect, and are affected by, each other. We have known that for a long, long time. But because it is intimidating to confront nature in all its complexity (and on account of differing interests, traditions, and training of scientists in various disciplines and subdisciplines), their development has historically been studied separately. But that is not to say we worked in anything like total isolation. From the beginning, for instance, studies of soil formation and evolution considered the effects of climate, organisms, geology, and topography on soils. The very concept of an ecosystem is based on the interaction of biota with their abiotic environment. And in the study of geomorphic landscapes, the role of changes in factors such as climate and vegetation cover has been prominent. Traditionally, though, we have not paid much attention to reciprocal interactions. Many studies, for example, have dealt with the effects of soils on vegetation or of vegetation on soils, but few studies have dealt with the simultaneous back and forth between the two.

The upshot is that ideas about the evolution of geomorphic, hydrological, ecological, and soil landscapes developed more or less separately. The separate threads have rarely, if ever, been irreconcilable or incompatible, but they have been mainly *independent* while the things they seek to explain are manifestly *interdependent*. As it should, that has begun to change. Understanding Earth's landscapes ultimately requires dealing with them on their own terms rather than *ceteris parabus*. Back when I was teaching physical geography, I taught the First Law of Geography as "everything is connected to everything else" (more on that below) and the Second Law as "all other things are never equal."

EVOLUTION

In science, the word *evolution* is used in several different ways. Sometimes it simply refers to development or change over time, not necessarily implying anything other than change that is systematic in some respect. In other cases, the term is used to contrast with words such as *development*, *succession*, or *progression* that imply deterministic progress along a single pathway toward a single end point. This use of evolution signifies the possibility of multiple pathways and outcomes and a potential role for chance and path dependence. A more precise notion of evolution also includes the role of selection processes. Evolution is used in the latter sense here, with other words such as development or succession applying to broader notions of change over time.

Many different areas of scholarship in the late nineteenth and early twentieth century were heavily influenced by the evolutionary ideas in Darwin's *On the Origin of the Species* (first edition, 1859). Scientists began looking for or interpreting their discoveries in terms of systematic changes over long spans of time. For instance, theories of topographic landscape development, ecological succession, and soil formation and change arose contemporaneously in this intellectual environment, with a common theme of systematic change with time. Some other important themes of evolutionary theory were less evident during this era, such as the roles of selection, chance, and historical contingency. The latter emerged in theories often viewed as rivals to Henry Cowles's and Frederick Clements's notions of succession, V. V. Dokuchaev's and Curtis Marbut's concepts of progression toward mature "climax" soils, and William Morris Davis's cycle of erosion (more on these below).

Admitting that it is difficult for an individual amid things to put them in proper context or to pinpoint what turns out to be revolutionary or transformative, one can at least testify about their own experience. For me, the past fifty years or so have been marked by the recognition of, and emphasis on, *interconnectedness*.

Landscapes are typified by dense webs of interactions in geographical space and through time, to the extent that changes to any aspect of the environment inevitably influences others. A principle arose that has been referred to as the First Law of Geography, the First Law of Ecology, and the First Law of Environmental Science, most famously originally articulated by K. E. F. Watt, Waldo Tobler, and Barry Commoner in the early 1970s.

EVERYTHING IS CONNECTED TO EVERYTHING ELSE

A corollary: you can't change part of a system (without also changing other parts or the system itself).

Of course, direct, evident, causal links do not exist between literally everything (and everything else). But everything *is* related to every other thing in that Earth systems are characterized by multiple components that are strongly interconnected, such that every component of the entire system is at least indirectly linked to every other. If you think of it (as I often do) as a mathematical graph (a way of representing networks of connections among components or locations), with the system components as nodes and the interrelationships as links between nodes, landscapes and ESS are *connected graphs*. That is, there is a path between any pair of nodes/components—sometimes direct, sometimes via intermediaries. In landscape science, straightforward cause-effect relationships between two things are rarely sufficient (though they are necessary pieces of the puzzle). Landscapes are more realistically thought of in terms of maps, webs, graphs, networks, matrices, flow charts, multiple equation systems, and other ways of representing and analyzing interconnected, mutually adjusting entities.

Additionally, all things are connected in that everything is wired into the same Earth system, so that somewhere along the way there exists a common history. For example, all organisms play a role in the global carbon/oxygen cycle, and all share common genetic ancestors.

Interrelationships among geophysical and geochemical phenomena, biota, landforms, and soils have long been recognized. One milestone was the Russian V. V. Dokuchaev's (1883) pioneering ideas of soils as the product of the combined influences of geology, climate, topography, and organisms. While Dokuchaev did not explicitly deal with reciprocal interactions, he did recognize that the factors of soil formation influence, and are influenced by, the other factors and the soil itself. Another Russian, V. I. Vernadsky, in 1926 came up with the concept of a globally interconnected biosphere and pioneered the study of biogeochemistry, whereby the biosphere, atmosphere, lithosphere, and hydrosphere are all linked in elemental cycles. The ecosystem notion (and possibly the term) predates A. G. Tansley, but his 1935 article introduced the concept to a broad scientific audience.

Studies tended to be one-way with respect to causality until relatively recently, however—for example, the effects of climate on plant geography or of topography on soil formation—rather than addressing interactions. Recognition that landforms, soils, biota, hydrology, and climate evolve and respond together, along with increasing emphasis on reciprocal interactions, led to the rise of several new subdisciplines. I have not tried to track down all the historical antecedents; the dates I give here are approximately when the subdiscipline began to coalesce and gain broader visibility within the scientific community. *Biogeomorphology* (or ecogeomorphology) emerged in the late 1980s and is concerned with the influence of landforms and geomorphic processes on the distribution and development of plants, animals, and microorganisms and with the effects of biota on Earth surface processes and landforms. *Ecohydrology* is at the interface of hydrology and ecology and, according to the journal of that name (first published in 2000), deals with interactions and feedbacks in space and time between ecological systems and the hydrological cycle. The journal *Geobiology* (which began in 2003) describes the subdiscipline as exploring relationships between life and Earth's physical and chemical environment. *Hydropedology*, which became a thing in the early 2000s, bridges pedology with soil physics and hydrology.

In recent decades, the terms *Earth system*, *climate*, *ecosystem*, and *critical zone science* have also become widely used to describe studies that transcend traditional disciplinary boundaries. Earth system scientists explicitly address the dense interconnections of the atmo-, hydro-, litho-, and biospheres and are especially concerned with these at broad (including global) scales and over timescales ranging from fluid dynamics to planetary evolution. *Climate science* is a term that encompasses not only climatology, paleoclimatology, and atmospheric sciences but also the study of climate impacts on, and feedbacks with, human and other biophysical systems. Ecosystem science is intended to signify that the study of ecosystems transcends ecology. The critical zone is defined (by the US National Science Foundation Critical Zone program) as "Earth's permeable near-surface layer . . . from the tops of the trees to the bottom of the groundwater." Critical zone science is an integrated approach to the study of rock, regolith,[2] soil, water, biota, and atmosphere interactions near Earth's surface.

These subdisciplines, terms, and concepts (along with more traditional ones such as biogeochemistry and geoecology) indicate recognition that no aspect of Earth's system(s) can be fully understood in isolation and that ESS are linked by constant internal and external feedbacks and reciprocal interactions.

So, start with the idea of *systematic change over time*. Then add notions of *selection*, *chance*, and *contingency*. Selection is

non-random; the most efficient, most resistant, best adapted entities are preferentially preserved and propagated. Selection is
also probabilistic (as opposed to deterministic) in that the most
efficient, resistant, or best adapted individual elements are not always, invariably, selected. Chance plays a role, including sheer
luck, as now and then "fitter" entities are destroyed and less-fit
ones perpetuated. Chance also incorporates disturbances, genetic mutations, and happenstance. Historical contingency reflects
the fact that present and future conditions depend partly on the
past. Next, mix in the interconnectedness whereby organisms,
climate, soils, landforms, hydrological, and other biogeochemical
fluxes (to name a few) affect, and are affected by, each other. Finally, stir in a context of constantly changing boundary conditions
(e.g., solar inputs, extraterrestrial impacts, plate tectonics, and human impacts). This recipe cooks up an approach illustrated by
Figure 1.1.

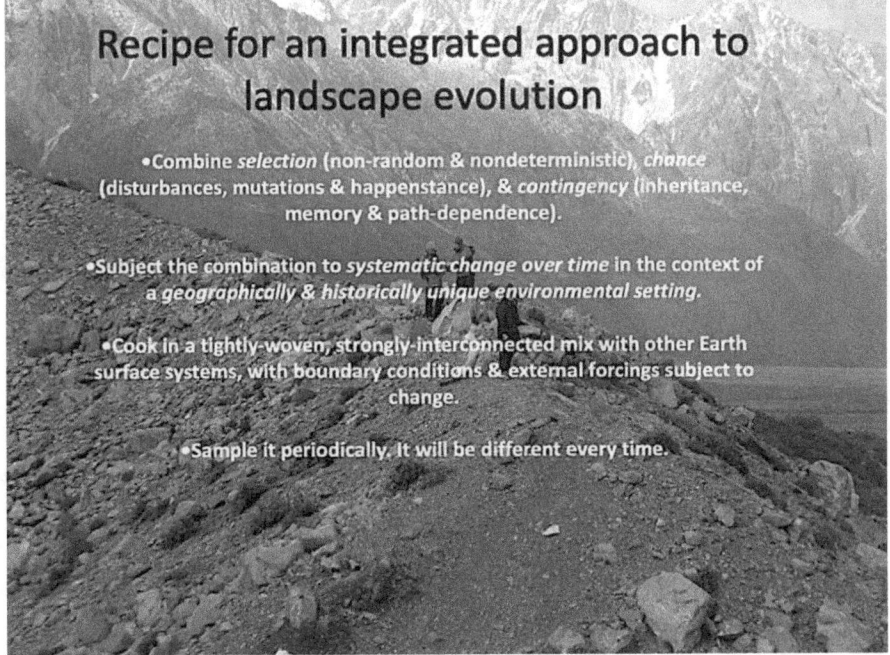

**Recipe for an integrated approach to
landscape evolution**

•Combine *selection* (non-random & nondeterministic), *chance*
(disturbances, mutations & happenstance), & *contingency* (inheritance,
memory & path-dependence).

•Subject the combination to *systematic change over time* in the context of
a *geographically & historically unique environmental setting*.

•Cook in a tightly-woven, strongly-interconnected mix with other Earth
surface systems, with boundary conditions & external forcings subject to
change.

•Sample it periodically. It will be different every time.

Figure 1.1. Integrated approach to landscape evolution. Background photo is a
moraine near Aoraki (Mt. Cook), New Zealand.

A COUPLE OF EXAMPLES

Landscape interactions occur at timescales ranging from the rates at which chemical reactions occur and sand grains bounce along a dune on a windy day to those of planetary evolution. These interactions are most evident, however, where they are neither too slow nor too fast to be readily observed.

Stream Channels

In teaching, I often used *hydraulic geometry* as an example of mutual adjustments in nature. Hydraulic geometry is the traditional term for the studies of interactions between channels and the flows they carry. Channels both shape and are shaped by those flows (Figure 1.2).

Figure 1.2. Channel of Board Camp Creek, Arkansas.

Let's start with the flow itself. Hydrologists, engineers, and geomorphologists make extensive use of *flow resistance equations*, all of which relate flow velocity to hydraulic radius, slope, and a roughness or friction factor. The first two promote higher velocity; the latter slows water down. Hydraulic radius is the cross-sectional area of flow divided by wetted perimeter (the two-dimensional length of the water-channel contact). If you think of this as mean depth of flow, you'll have the right idea, and in many cases mean depth and hydraulic radius are nearly equal. Slope refers to energy grade slope, the change in hydraulic head over distance. If you think of this as the slope gradient of the water surface, you'll again have the right idea. Friction factors or roughness coefficients reflect slowing of the flow due to friction with the bed and banks, influenced by the geometric irregularity of the channel, obstructions, debris, and vegetation.

The flow resistance equations reflect the basic physics of water flow and can be rearranged, with no loss of legitimacy, to solve for hydraulic radius, slope, or friction factor as well as velocity. In other words, each of these fundamental hydraulic variables is a function of the others, as shown in the upper part of Figure 1.3. The lower part of Figure 1.3 shows interrelationships from what is known as the flow continuity equation, whereby discharge (flow rate) is the product of velocity and cross-sectional area, which is in turn the product of width and depth. Of course, this is also a reciprocal relationship, as the amount of water in the channel and how fast it is moving is a function of how much water is supplied to it (i.e., discharge).

Now we figure in the morphology of the channel itself. Slope is partially imposed by the topographic gradient of the channel, and width and depth of flow are strongly constrained by the width and depth of the channel. Conversely, channel geomorphology is largely a product of the erosional, depositional, and sediment transport effects of the flows that pass through.

Lots of things live in and adjacent to streams. The geomorphology and the flow characteristics determine the habitat suitability for various organisms, and those organisms in turn influence channel morphology and flow. Think about plants and wood and other plant debris in channels for a start, but it is only a start. Algae

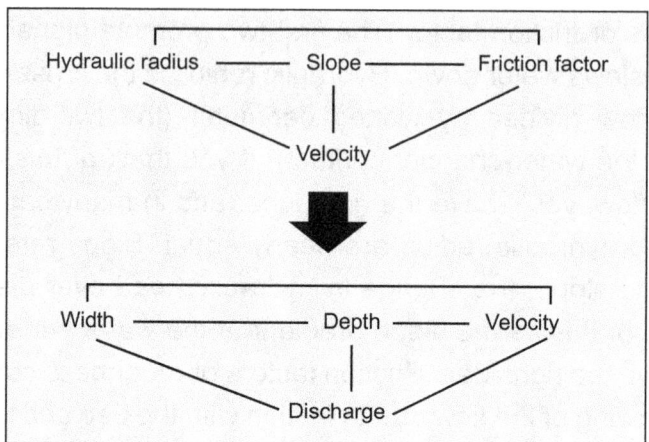

Figure 1.3. Mutual adjustments in stream flow, shown as graphs. Top, interactions implied and reflected by flow resistance equations. Bottom, interactions implied and reflected by flow continuity.

and microbes bind sediment, exude substances into water and sediment, and influence the roughness (or smoothness). Fish, crayfish, and other fauna dig spawning holes, burrows, and tunnels, and these and many other critters stir up the bottom looking to eat other critters. Then there are beavers and their dams, bars, and shoals composed of shells, effects of roots on stream banks, and so on.

Figure 1.4 illustrates the mutual adjustments among geomorphology, flow, and biota. Like many other landscapes (or waterscapes), stream channels are complex, mutually adjusting systems.

As any fluvial geomorphology text will reveal, there exist countless varieties of streams and rivers, which occur within an endless combination of climate and geological settings, biological and ecological frameworks, human impacts, ages, stages, and histories. Yet within this bewildering and wonderful variety, and with many critical variables that both affect and are affected by each other, certain regularities of patterns, forms, and relationships exist. Why? It's not like a river or rock or raindrop can strive toward some goal, and it's not like general principles are the only thing at work here.

Rivers work in mysterious ways.

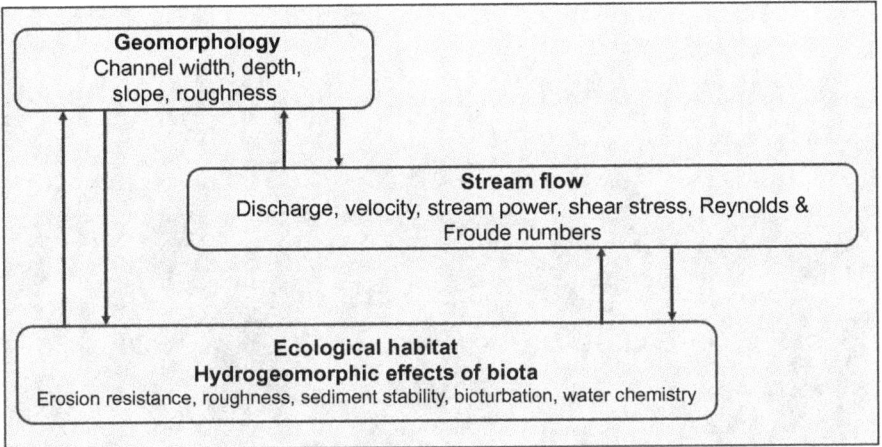

Figure 1.4. Mutual interactions among geomorphology, flow, and habitat in stream channels.

Meet the Beetles

In 2018, Pavel Šamonil, Lukasz Pawlik, and I were studying forest biogeomorphology in old-growth forests in the Czech Republic. In our travels from one site to the next, we began talking about the many Norway spruce (*Picea abies*) stands we had seen damaged or destroyed by the spruce bark beetle (*Ips typographus*; Figure 1.5). Given what we (and many others) had been finding in recent years about the profound impacts of trees on landforms and soils, an obvious question arose. If trees have major impacts on landforms and soils, and bark beetles have major impacts on trees, then shouldn't we expect bark beetle outbreaks to have important effects on landforms and soils? There is extensive literature on the biological, ecological, and economic aspects of spruce bark beetles (SBBs) but very little on their indirect impacts via their tree-eating. Those discussions in the backwoods and on the backroads and in small-town cafes of Czechia eventually evolved into an article called "Indirect Biogeomorphic and Soil Evolutionary Effects of Spruce Bark Beetle," published in *Global and Planetary Change* in 2020. It is another good illustration—though I leave the scientific details to that article—of the interconnections in ESS.

Figure 1.5. Norway spruce stand destroyed by spruce bark beetles, Sumava National Park, Czech Republic.

Figure 1.6 walks through *some* of the chain of interactions involved in addressing this question. This example is typical in another way as well: we have a pretty good idea of how the individual causal links work, but their net effect and how all the pieces fit and work together are not yet well known.

Bark beetles kill trees. But even that has mutual interactions because the SBBs are attracted to recently deceased or stressed trees and then spread to healthy trees. Norway spruce typically have shallow, wide-spreading root systems that help them adapt to the wet sites where they are often found. As such, they are more prone to uprooting than most other trees. Uprooting not only disturbs the soil as the toppled tree's roots rip up the material it is rooted in but also leaves behind a characteristic pit (where the roots were ripped out) and mound where the rootwad soil eventually settles. This hummocky pit-mound topography strongly effects runoff, directing more of it to the subsurface as runoff sheds off the mounds and goes into the pits. Even if the tree is not uprooted, a dead

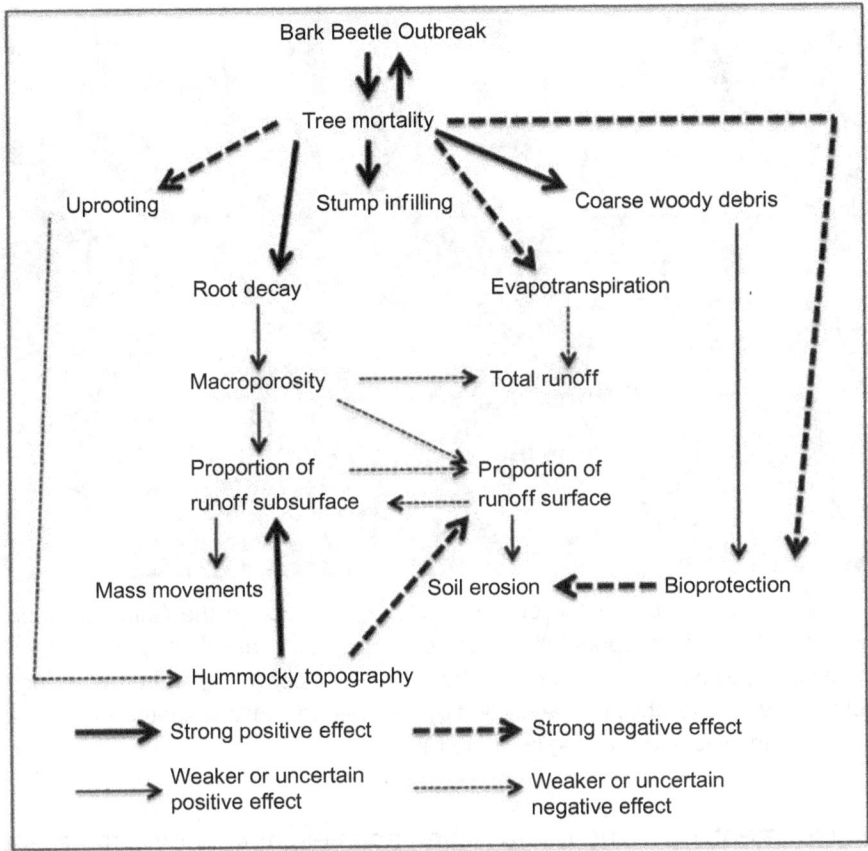

Figure 1.6. Spruce bark beetle impact on landforms and soils. (Reproduced with permission from P. Šamonil, J. D. Phillips, and Ł. Pawlik, "Indirect Biogeomorphic and Soil Evolutionary Effects of Spruce Bark Beetle," Global and Planetary Change 195 (2020): 103,317)

tree eventually leaves a stump, which leaves a runoff-gathering depression as the wood decomposes (Figure 1.7).

Tree roots provide preferential paths for infiltration and subsurface flow along the roots when the tree lives and along root channels when the dead root decays. A mass death of trees, not uncommon after a beetle outbreak, reduces the amount of water transpired by trees while increasing the laterally oriented subsurface flow along the root channels. Because the rooting habits and uprooting tendencies of *Picea abies* differ from other trees in the region (we focused on central Europe), if SBBs result in

Figure 1.7. Microtopographic effects of Norway spruce in the Sumava Mountains. Clockwise from upper left: hummocky topography associated with mounding by living trees, pits, and mounds from past tree uprooting and downed logs; hummocky topography in a stand killed by bark beetles; pit associated with rotting of stump and roots; mounding by a live tree.

replacement of spruce with other species, these dynamics are changed.

A tree-killing event also adds a lot of dead wood to the surface at once. This provides some bioprotective effects from erosion, but different from those of living trees, and adds a big influx of cellulose-rich organic matter to the soil.

The bottom line is that while we may be confident that the SBB has significant indirect impacts on soils, landforms, and hydrology, the cumulative impact and how various effects offset or reinforce each other are still unclear. Clearing it up will be *relatively* simple (but by no means easy) with respect to how changes occur at specific sites and with respect to the rates and processes of specific phenomena, such as the decomposition of dead trees and root systems under various environmental conditions. Putting all those pieces together into a coherent story will be far more complicated. And determining how (or whether) this fits into some broader narrative takes us into the realm of mysterious ways.

COEVOLUTION

Coevolution occurs when entities mutually evolve, affecting and affected by each other. Coevolution in biology involves reciprocal interactions among taxa that select for or against individual organisms. Among other things, these may be in the form of predator-prey "arms races" or mutualistic interactions such as relationships between flowering plants and pollinators or between grasses and grazing herbivores.

The coevolution concept has been extended by paleoecologists and paleopedologists to include interrelationships among biota and abiotic factors as well as interactions among organisms. *Abiotic* is used broadly here for factors that are not organisms and that are strongly influenced by abiotic phenomena *sensu stricto*. This qualifier is necessary because it could be argued that few things in landscapes are not influenced at least indirectly by biota. Some inputs to ESS such as gravity, plate tectonics, or solar radiation inputs are indeed purely abiotic. The impurely broad use of abiotic here includes environmental factors such as atmospheric, water, and soil chemistry that are affected by both biota and geochemical/geophysical factors as well as strictly abiotic factors.

A good example of biotic and abiotic coevolution involves grassland environments. Back in 1883, Dokuchaev showed that regardless of the geological parent material, grassland soils have properties attributable to vegetation, organic matter, and earthworms. Later, others (particularly Gregory Retallack of the University of Oregon) showed that grassland soils have coevolved with grasses themselves, and with grazing mammals, over the past forty million years (Ma), with (of course) interactive effects of landforms, soils, and biota on each other. This coevolution was revealed in the stratigraphic record of lower Eocene (56 to 34 Ma) rocks. Other studies demonstrated complex coevolutionary interactions among climate, soils, herbivore densities, and grass traits. Not unexpectedly, paleoenvironmental data show that changes in Earth's atmospheric composition are reflected in both ancient paleosols and in paleovegetation. Atmospheric chemistry is itself greatly influenced by plants and soils, so this is a clear example of coevolution.[3]

Many other examples of biotic/abiotic coevolution exist and will be touched on later.

SELECTION

Selection is most familiar in the biological realm. Selective breeding by humans, for instance, turned wolves into the incredible variety of dog breeds now extant and transformed wild South American plants that could be easily mistaken for weeds by non-botanists into maize. *Natural selection* was coined by Darwin to contrast it with such human (artificial) selection. Mechanisms and nuances of natural selection continue to be debated and investigated, as they have been since at least 1859. The undisputed core concept, however, is that organisms with unfavorable traits are selected against, while favorable traits are selected for. Favorable in this case means those that increase the odds of survival and reproduction. As we know all too well in human affairs, the virtuous, talented, and deserving are not always rewarded, while ignorant, vile, and venal people sometimes are. Natural selection does a far better job of favoring the better adapted, but it is still probabilistic and imperfect.

Some argue that anything other than Darwinian natural selection, *sensu stricto*, or at least directly and closely analogous to it, is not really selection. Pay them no mind. Such arguments are of no interest here and are semantic at best. Natural selection in biological evolution is a specific case of preferential survival and propagation of some entities and preferential loss or failure of others. Other types of such preferential survival and propagation can be equally important and interesting.

Various strains of so-called generalized or universal Darwinism, applied in business, social sciences, and metaphysics, seek to apply concepts of variation, selection, and heredity to a wide range of subjects. Universal and generalized Darwinism is, to varying degrees, argument by analogy. That is, if something is somehow analogous to biological evolution, then generalized Darwinism applies. This is unnecessary for this book, however. Arguments by analogy are unnecessary because selection principles are *directly* applicable to a range of geophysical, chemical, and system-level phenomena. More about this in Chapter 6.

GOAL FUNCTIONS AND EMERGENCE

I use the term *goal function* to refer to hypotheses that ESS seek to maximize, minimize, or optimize something, usually referring to some aspect of mass/energy flux and storage. These proposals imply that these are somehow goals or aims of ESS. Of course, landforms, soils, and ecological systems have no intentionality— nor do flows of water, transformations of energy in food webs, or other processes. Landscapes have no ability to literally adopt goal functions, and no such intentionality is directly suggested, only im- plied. The hypotheses I refer to as *goal functions* are sometimes proposed to represent scientific laws and sometimes a rhetorical, metaphorical tool to describe ESS development.

Some goal functions are general, as with various hypothe- sized energy-related goal functions in ecosystems. Others are specific, including some related to flow-sediment-channel inter- actions in streams, where maximization or minimization of vari- ous hydraulic parameters (sometimes called extremal hypotheses) are proposed. Still others, such as the notions of progression toward climax soils and vegetation communities, are quite general.

Phenomena that arise spontaneously and are not inevitable are called *emergent*. Emergence may, and often does, occur due to general laws or principles, but the emergent phenom- ena are not *required* by those laws. Outcomes once viewed as manifestations of goal functions are increasingly seen as emer- gent. Because emergence is a much simpler explanation, it is to be preferred over goal functions when both are capable of accounting for observations. Andrei Lapenis of the State University of New York at Albany, for instance, showed in 2002 that apparent "Gaian" evolution at the level of ecosystems or biospheres could emerge due to biogeochemical selection, whereby faster or more efficient material cycling is advantageous. Emergent products of basic principles of gradient and resistance selection, I argued in 2011, can explain landform evolution better and more simply than purported goal functions of steady-state equilibrium.[4]

LAWS, PLACE, HISTORY

An approach to the study of landscapes that recognizes the combined, interacting effects of general laws and local geographically and historically contingent factors is the law–place–history (LPH) framework (Figure 1.8). LPH can serve as a pedagogic device—I used it in my University of Kentucky classes for years—or as an instrumental tool for research and interpretation.

Law refers to laws per se (for example, the conservation of mass, energy, and momentum) but also to other generalizations that hold no matter the time or location. These are linked to the *ceteris parabus* principle. All other things being equal, soil erosion decreases or increases with greater or less vegetation cover, for example. Some laws are quantitative and specific (for instance, the equations for dissolution of calcium carbonate or wave dynamics). Others are general and qualitative, such as all other things being equal, wind erosion is inversely related to soil moisture.

Place incorporates the geographic framework for landscapes. Place is the "other things" that are *not* equal—that is, the environmental context in which laws operate. Place factors may be general, such as climate, land use, or geology. They can also be specific, such as particular metrics of climate (e.g., soil temperature or heating degree days) or measurements of things such as bedrock fracture spacing or percentages of different land covers.

History factors comprise the path-dependent, historically contingent aspects of landscapes. They may include the timing of events or their sequence. History also includes ESS sensitivity to initial conditions, time available for system evolution, previous evolutionary pathways, disturbance histories, and stages of development. Coastal landscape morphology and hydrological and ecological dynamics, for instance, are often strongly affected by the sequence and timing of storm events, elapsed time since the last storm, and vegetation history such as successional stages and introduced species. Landscape evolution is also influenced by history in the form of inherited environmental controls (for example, when ancient geological structures influence modern weathering and erosion).

In a particular landscape or context, it may be difficult to designate a given factor as strictly a law, place, or history factor. What to do then? Don't worry about it. Pick one or put it in two or three categories. The important idea is not to pigeonhole everything as L, P, or H but to instead recognize that all three types of factors are important.

TREE: TRANSFORMATIVE, RECURSIVE, EMERGENT EVOLUTION

TREE is an acronym summarizing the nature of landscape development and change as *transformative*, *recursive*, *emergent* *evolution*. Transformative indicates that landscape change often involves not just incremental or quantitative changes but also wholesale changes in ecosystems, landforms, soils, and hydrological systems and their interconnections. These can often be described using terms such as *state changes*, *regime shifts*, or *metamorphoses*. These transformations may be relatively abrupt, or cumulative over long time periods, and may be driven by external factors, internal interactions, or both. Recursive refers to the coevolution of landforms, ecosystems, soils, and hydrologic

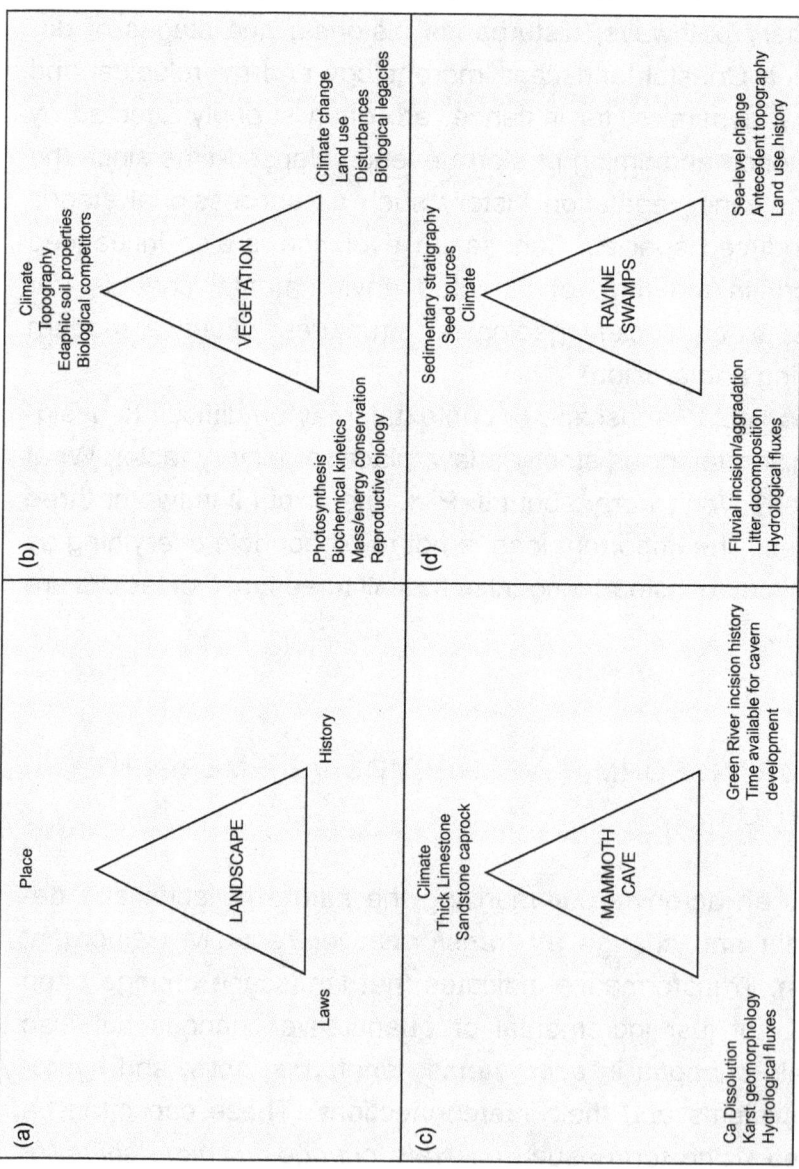

Figure 1.8. The LPH framework. The most general form (A); as applied to explain vegetation community composition at a given location (B); as I have used it to explain the formation of the world's longest cave in Kentucky (C); and as applied to a particular feature found in the outer coastal plain of North Carolina (D). In B–D the indicated law, place, and history factors are an incomplete list given as examples.

regimes. The landscape components affect, and are affected by, each other so that development is characterized by ongoing mutual adjustments. Emergent means that while patterns, structures, and networks of relationships develop, they are not always deterministic and are not guided or preordained. These characteristics emerge due to the interactions within the landscape in the context of its boundary conditions. Evolution reminds us that landscape development is historically and geographically contingent (path dependent), affected by selection, influenced by chance events, and probabilistic.

In this section, three examples of TREE are briefly reviewed, representing different environmental settings and timescales.

WHITE DESERT, EGYPT

Egypt's western desert (west of the River Nile) includes the White Desert, a karst massif known for its spectacular landforms and white rock outcrops and features. The landforms are primarily karst in origin—but karst requires humid conditions, whereas the White Desert is not just dry but also hyperarid. The region thus represents an example of landscape transformation over long geological timescales that includes strong elements of inheritance and historical contingency.[5]

Sixteen fields of paleokarst (originally formed by karst processes that are now no longer active) landforms have been recorded in Bahariya-Farafra Oasis areas of the western desert (Figure 1.9.) The White Desert includes large karstified late-Cretaceous carbonate platforms, which were subject to dissolution and other karstic processes from the late Cretaceous to the late Eocene, guided by structural lineaments comprising E-W- and N-S-trending faults, fractures, and joints. After the karst modification up to the late Eocene, there were phases of uplift and tectonic/climate change favoring karstification during the Miocene. Wind and water action affected by local structures modified the karst features.[6,7]

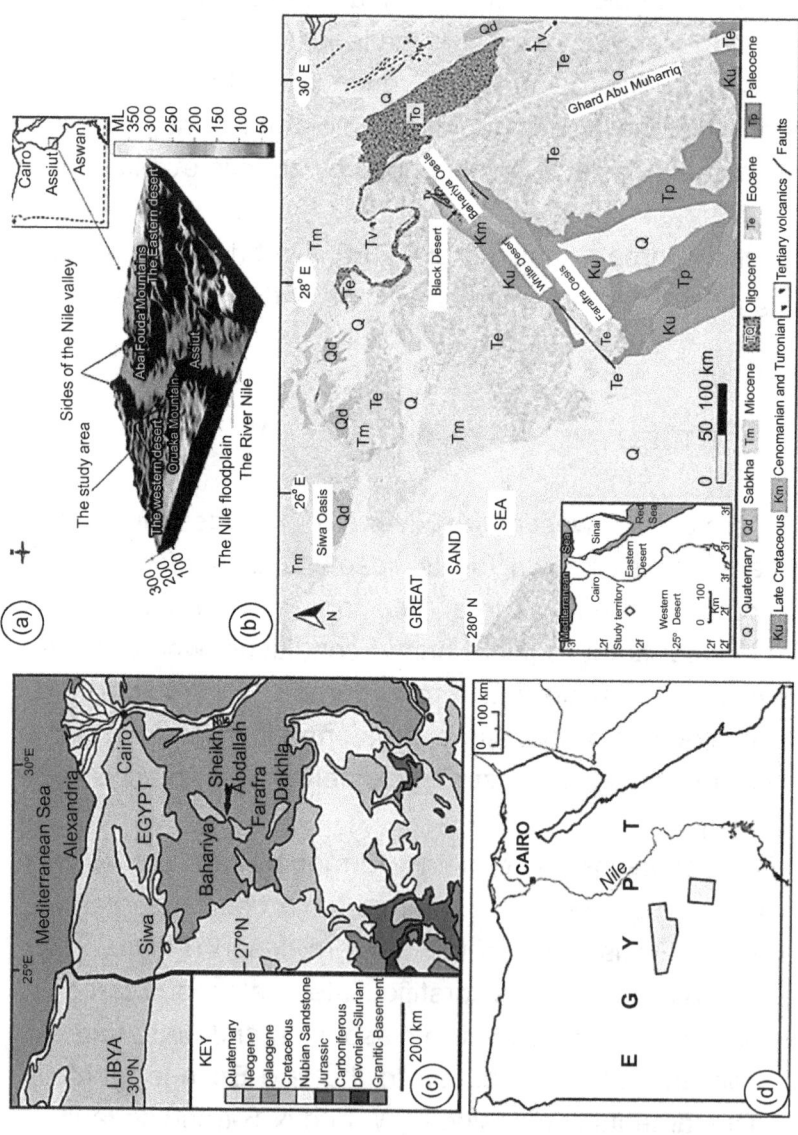

Figure 1.9. (A) Satellite-derived image showing setting of White Desert National Park. (B) Geologic survey of Egypt map of surficial geology. (C) Geological map of Egypt. (D) Exposures of karstified carbonate rocks exposed in the western desert. Reproduced under a CC-BY 4.0 license from E. A. Khalaf, "Karst Heritage as a Tourist Attraction: A Case Study in the White Desert National Park, Western Desert, Egypt," Geoheritage 14 (2022): 94.

Karst landforms include poljes, which are large structurally controlled depressions. In the White Desert area, they range from 980 to 1800 km^2 in area. Some are temporarily transformed into lakes by floodwaters (including groundwater rising from filled depressions below). Smaller karst depressions include dolines (sinkholes or cenotes) and various sinks and swallets. Karst pavements are large low-relief plains with various dissolutional forms such as karren and solution pits. Some areas of the pavements are covered by rounded siliceous chalky limestone concretions sometimes referred to as cannon balls or melon fields. These signify local-scale microbially induced weathering during advanced stages of the karstification processes.

Some of the most striking landforms are mushroom karst, with 1 to 5 m vertical stems topped by irregularly shaped necks and caps, often bowl-shaped or bulbous (Figure 1.10). The shapes result from lithological variations—the stems are mainly soft calcareous sandstones, while caps are mostly hard siliceous limestone. Also common are inselberg-like rock features shaped by differential dissolutional weathering and sculpted by wind erosion. These represent more resistant rocks, preserved by resistance selection, rising above the karst pavements. Tower, bridge, and cone karst features also occur.

Smaller-scale features include speleothems originally formed by mineral precipitation in caves and exhumed cave walls featuring various types of weathering and precipitation features.[8] Well-preserved fossil mangrove plant roots are present at some sites (Figure 1.11).

The evolution of the White Desert landscape from a humid, karst environment to a hyperarid desert dominated by relict, paleokarst forms occurred between the late Eocene (which ended 33.9 Ma) and the mid-Holocene, where radiocarbon dates and archeological data indicate depopulation of the region corresponding with a shift from more humid to hyperarid conditions.[9]

The transformation involved intermediate states, was influenced by a series of tectonic episodes and climate changes (among

Figure 1.10. Examples of features in White Desert National Park that originated as karst landforms and subsequently modified by aeolian erosion. (A) Mushroom rock. (B) Cannonball concretions with karst pavement in foreground. (C) Inselbergs formed from paleokarst cones. (D) Wave rocks—wind-abraded features modified from resistant carbonate outcrops.

other things), and was more complex than an Eocene karst vs. modern comparison suggests. Nevertheless, such a comparison does reveal TREE (Table 1.1). Transformative aspects of the changes are obvious. Recursivity among hydrological, biological, geomorphological, and pedological processes always exists in active karst landscapes. Such mutual adjustments and reciprocal interactions are less obvious and intense in the modern drylands, but they still occur. This is most evident in the concentration of resources and ecosystem services in oases. In the modern hyperarid

Figure 1.11. (A) Paleokarst cavity used as nest site by sooty falcon (*Falco concolor*). (B) Fossilized mangrove roots. (C) Oasis resulting from resource concentration in the post-Eocene landscape.

Table 1.1. SUMMARY COMPARISON OF THE WHITE DESERT NATIONAL PARK AREA AT THE TIME OF KARSTIFICATION AND THE MODERN HYPERARID DESERT

	Karst Phase	**Hyperarid Phase**
Hydrological Regime	Combination of ground-water and surface flows interconnecting among solutional depressions, subsurface conduits and caves, and surface runoff	Little runoff; occasional filling of paleokarst depressions during short-term flood events; isolated springs (oases) fed by groundwater
Chief Landforms	Active karst features	Relict paleokarst features, with arid/aeolian landforms superimposed
Major Erosion Process	Solutional denudation	Aeolian abrasion

	Karst Phase	Hyperarid Phase
Major Weathering Process	Dissolution of carbonate rocks	Salt weathering (hydration)
Biome/ Ecosystems	Subtropical forest	Hyperarid desert
Soils	?	Exposed bedrock, Eutric Leptosols, Gypsic Solonchaks, Calcaric Cambisols, Calcaric Arenosols[†]
Biological Influences of Landforms	Carbonate weathering driven by CO_2 production from respiration and organic matter decomposition; formation of organic acids; biological activity in rock joints; biogeomorphic ecosystem engineering	Minimal (other than human agency)
Landform Influences on Biota	Formation & modification of habitats; weathering role in biogeochemical cycling	Specialized habitats for desert organisms; concentration of resources at oases
Inheritance	Carbonate sediments & small-scale features (e.g., fossil mangrove roots) inherited from carbonate platform formation	Paleokarst landforms

[†]World reference base soil classification; soil information from K. Kinderman et al., "Palaeoenvironment and Holocene Land Use of Djara, Western Desert of Egypt," *Quaternary Science Reviews* 25 (2006): 1619–1637.

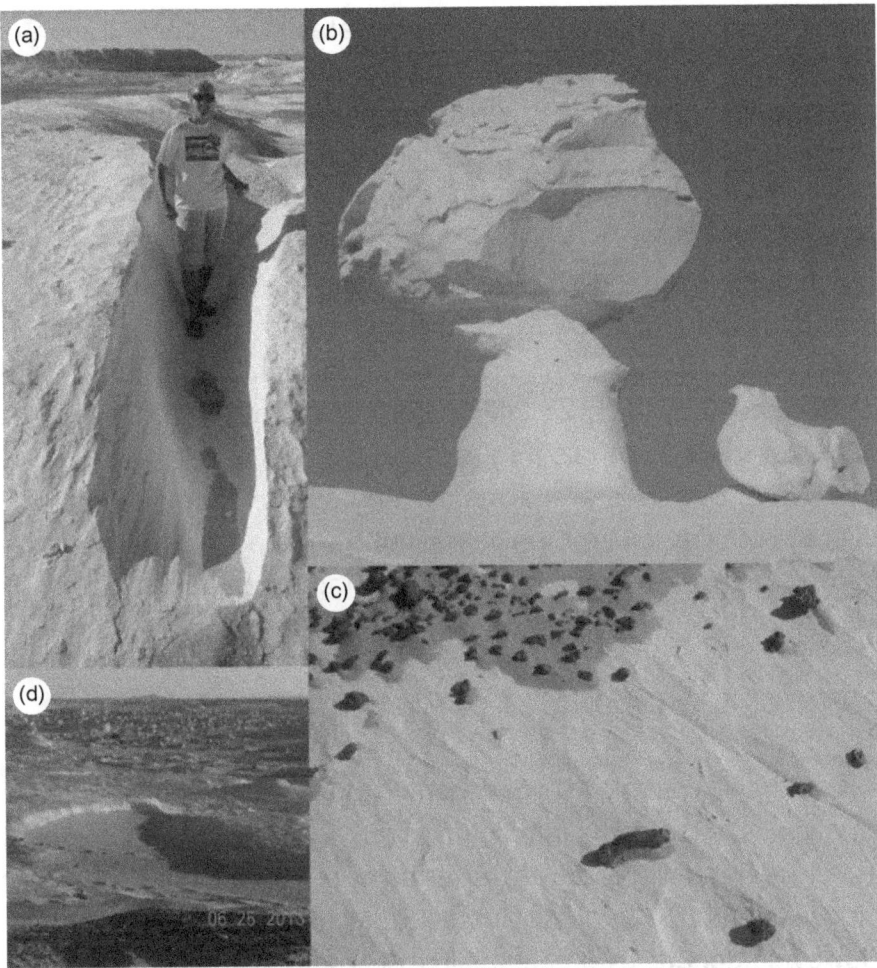

Figure 1.12. Evidence of efficacy of aeolian abrasion in the White Desert. (A) Ventifact that is most likely a modified paleokarst feature. (B) Mushroom rock formed by differential abrasion of calcareous sandstone (stem) and hard limestone (cap). (C) Karst pavement with black volcanic gravel that can only have been transported by wind. (D) Mixture of wind-modified paleokarst and aeolian sand.

White Desert, precipitation is minimal (mean of 6 mm yr^{-1}), and evaporation (where there is any water to evaporate) is seventeen times higher. The only source of water is groundwater, which comes from a deep sandstone aquifer that pushes groundwater up into the Eocene carbonates. Other than by humans drilling wells, this can only reach the surface where the paleokarst allows it or in paleokarst depressions (poljes) floored by older rocks due to the erosional removal of the carbonates. The concentration of the critical water resource influences or determines biogeographic patterns and human activities. Though the groundwater aquifer is partly replenished from recharge areas outside the White Desert region, the age of groundwater in the oases is estimated to be 5 to 20 ka.[10]

A notable aspect of the transition is the shift (once again, comparing the late Eocene with the present) from a domination of karst processes to aeolian. Wind abrasion is exceptional in the region, as evidenced by the sculpted landforms, and by the rare transport of gravel-sized materials (Figure 1.12).

The White Desert is a spectacular and scenic example of TREE over about thirty-four million years, but comparably profound transformations over millions of years are common on our planet.

SIERRA NEVADA DE SANTA MARTA, COLOMBIA

A common example of landscape transformation over shorter timescales is the conversion of tropical moist forests to savannas. This is commonly discussed in terms of recent and contemporary modifications due to direct human impacts such as deforestation. However, debate over the origins of tropical savannas and the role of anthropogenic factors dates back at least to Alexander von Humboldt in 1818. Here we focus on the work of Jaime Cavelier (University of the Andes, Colombia) and T. M. Aide (University of Puerto Rico), along with C. Santos, A. M. Eusse, and J. M. Dupuy. They studied the "savannization" of moist forests in the Sierra Nevada de Marta in the upper Rio Rancheria watershed.[11]

In their study, region savanna vegetation is interspersed with moist forests (Figure 1.13). The former is dominated by native grasses and dispersed trees that are also native to the neotropics. Though moist forest patches are more common on lower slopes and in valley bottoms, they also occur on mid and upper slopes and occasionally on ridges. These forest patches are not gallery

forests as are found throughout the neotropics but are the result of fragmentation of a once-continuous forest. Cavelier et al.
in 1998, indeed, ascribed the transformation to anthropogenic factors, but not recent human agency. They attribute initiation of the
transformation to Amerindian use of fire to enable cultivation of
maize in pre-Columbian times (Figure 1.14), followed by the introduction of cattle by Spaniards ca. 1530 and the continued use
of fire to maintain grazing land.

Figure 1.13. Savanna/forest boundary. GoogleEarth[TM] image.

The savannization is more than just a shift in vegetation cover.
In comparing savanna and forest patches on the same slopes,
they found that A and B soil horizons (topsoil and subsoil, respectively, typically amounting to about 0.5 m in thickness in the region)
were absent under the savanna, with no corresponding changes in
underlying bedrock or soil moisture status, though in general the
savannas are more arid. This indicates the strong, recursive interrelationships among vegetation, soils, landforms (via soil erosion),
and hydrology.

Figure 1.14. Cuidad Perdidas pre-Columbian archeological site in moist forest, Sierra Nevada de Santa Marta. Photo credit: Raphael Chay, reproduced under a CC-BY 3.0 license from Wikimedia Commons.

Savannas do occur in the absence of human activity in the region, as evidenced by paleoecological data, as well as their occurrence in apparently undisturbed locations. Though there is no evidence that natural dry savannas can occur within moist forests except on poor soils, this does suggest that naturally occurring fire and localized erosion could result in savannization without humans. The paleoecological record also shows forest-savanna transitions in both directions occurring in response to climate change, particularly modifications of moisture and fire regimes. This suggests an alternative stable-state situation, though severe erosion that results in persistent soil fertility loss may prevent transitions (back) to the moist forest.[12]

Human actions, whether more intense post-nineteenth-century modifications or more modest pre-Columbian ones, are clearly responsible for deliberate (as opposed to emergent) transitions. However, a broader perspective suggests that these can and did occur in an emergent fashion, making this an example of TREE.

LOWER NECHES RIVER, TEXAS

This example is based on the lower Neches River of southeast Texas, downstream of its confluence with the Angelina River and above its deltaic zone. We focus on landscape transformations associated with avulsions (a shift or relocation of a river channel) and formation of multiple-channel systems.[13]

Landforms

Two types of pre-Holocene landforms influence avulsions in the lower Neches, reflecting historical contingency. Quaternary sea level changes have formed three sets of terraces within the river valley, as base level reversals drove alternative episodes of aggradation and floodplain building, followed by downcutting (the current regime is rising sea level and aggradation). The other main inherited features are large paleomeander scars and depressions within the valley bottom, associated with higher river flows in the late Pleistocene. The paleomeander features are readily distinguished from more recent meander scars and oxbows by their far wider (paleo-) channels and much greater amplitudes (Figure 1.15). They seem to have formed about 18 to 10 ka, when paleoclimate evidence suggests wetter conditions.[14]

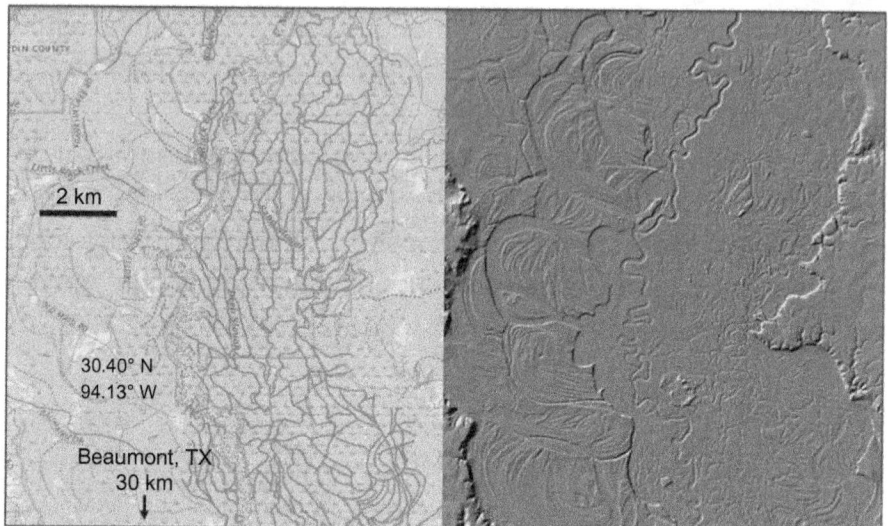

Figure 1.15. A portion of the lower Neches River valley. Relief map on right of the same area as that on the left shows the large paleomeander scars on the west side of the river and the complex topography of the floodplain. Elevation of the Pleistocene alluvial terraces including the paleomeanders is about 8 to 14 meters above sea level (masl). Floodplain elevations are < 10 masl, with a total relief of < 3 m.

Avulsions in alluvial rivers are an example of landscape change often driven by internal dynamics of the fluvial system, though they can certainly be affected by external factors. They are best understood as a setup-and-trigger phenomenon. Setup conditions include channel aggradation and a potential cross-valley slope advantage—that is, a potentially steeper path for the flow to follow were it to break out of its existing channel. The latter is typically associated with former channels (some rivers avulse a lot!) or depressions on the floodplain (including paleomeander features). These are necessary, but not sufficient, conditions for an avulsion, though in cases of so-called superelevation (the channel bed aggrades to the point that it is higher than the floodplain on the other side of the natural levee) an avulsion at some point is well-nigh inevitable.

Triggers are factors that force or allow flow to cut through the natural levee during near-banktop stages or floods. These include logjams or large sediment accumulations or low or weak spots in the levee that let flow through. When flow breaches the natural levee, it's called a crevasse, and sometimes the flow simply spreads out and deposits sediment in a fan-shaped crevasse splay. If the crevasse flow is concentrated and forceful enough, it can cut a channel, which naturally works its way along the steepest path available. If the new channel captures enough flow to persist after the high flow and crevasse event are over, you have an avulsion. Sometimes the channel is entirely shifted; sometimes both the new and old channel persist. If the channels rejoin downstream, you have a multichannel system—an anabranching or anastomosing system. Anastomosing channels require avulsions, and the Neches is full of them, both on the main active river channel(s) and on smaller floodplain channel systems (Figure 1.15).

Avulsions create a variety of channel types, applicable to both the Neches River and its tributaries. These include the main modern channel and active and semiactive subchannels and abandoned channels and meanders, which represent both the river and its tributaries.

Abandoned channel waterbodies (ACWBs) are sloughs and oxbows that do not normally convey flow but are persistently inundated. Those that eventually infill mainly with sand represent favorable locations for future avulsions, providing both local depressions and erodible sediment. Those that infill with clayey material are more resistant and may inhibit future avulsions.

Active channels have banktop ridges (natural levees) associated with local deposition as flows go overbank. As channels migrate laterally, former levees are often preserved as alluvial ridges, with intervening swales. Backswamps are lower floodplain areas behind the levees, with or without ridges and swales.

Superimposed on floodplain surfaces are minor topographic features associated with local scour and deposition, organic matter and woody debris accumulations, uneven sediment deposition, tree uprooting, and faunal effects (e.g., alligator trails and wallows, feral hog foraging, ant mounds). These may be particularly important in this environment where elevation differences of a few centimeters may be hydrologically and ecologically significant.

Hydrology

The landform variability is associated, and mutually adjusts, with a mosaic of hydrologic regimes, including permanently and intermittently flowing channels, with the latter ranging from those that normally convey flow but occasionally dry up during droughts to those that are usually dry but occasionally flow during floods (Figure 1.16). ACWBs exhibit six hydrologically distinct states:

Figure 1.16. A floodplain depression (top) and an anabranch channel (bottom) of the lower Neches River.

flow-through, flood channels, fill-and-spill, fill-and-drain, tributary fed or occupied, and disconnected. These differences are critical to channel-floodplain connectivity.[15]

Within the valley bottom, landforms and habitats are distin-guished by their *hydroperiod*, a general term for the frequency and duration of inundation. Flooded areas are classified as occasion-ally, seasonally, frequently, or permanently flooded. Ponded sites are designated as frequently or occasionally ponded. There also exist some rarely flooded or ponded sites.

These can be further characterized according to whether the ma-jor water source is the Neches River, tributaries, or local runoff and water table rise.

Soil hydrology also varies, mainly with respect to the typical height of the water table over the course of a year and the soil's drainage capacity, which is controlled primarily by saturated hy-draulic conductivity. Soil drainage classes range from excessively drained (sandy and gravelly soils that do not retain water) to very poorly drained. These are linked to soil texture (mainly sand vs. clay) and water table elevations.

Soils

Table 1.2 shows the seventeen recognized soil series mapped in the lower Neches River bottomlands, including some soils associated with Pleistocene alluvial terraces on lower levels within the valley bottom. Many occur as "islands" isolated by Holocene avulsions.

The main factors distinguishing the soil types are (or are closely related to) the dominant texture, the prevalence of low-activity vs. smectitic[16] clays, and soil drainage. Sandier soils form in former channel, point bar, crevasse splay, and natural levee de-posits. Clay-rich soils develop mainly where flow was impeded or water was ponded, providing time for clay-sized materials, which have very low settling velocities, to be deposited, such as abandoned channels, backswamps, and floodplain depressions.

Table 1.2. SOILS OF THE LOWER NECHES RIVER. TEXTURE INDICATES THE DOMINANT TEXTURE.

Series	Taxonomy	Texture	Setting	Interpretation
Belrose	Oxyaquic Paleudult	Sand	Pleistocene terrace risers affected by seasonal high water table	Well-developed sandy soil on former floodplain
Caneyhead	Typic Glossaqualf	Sand	Floodplain depressions	Bisequal; new deposition on older alluvium
Cowmarsh	Typic Fluvaquent	Clay	Holocene ACWBs	Late stages of channel infill with fine overbank material and organic matter; dominant upper Neches[†] source for underlying material
Estes	Aeric Dystraquert	Clay[*]	Floodplain	Recent clayey alluvium from Neches
Hatliff	Fluventic Dystrudept	Sand	Natural levees & point bars on floodplains	Recent sandy alluvium
Iuka	Aquic Udifluvent	Sand	Stratified loamy and sandy alluvial sediments on floodplains	Recent sandy alluvium

Series	Taxonomy	Texture	Setting	Interpretation
Iulus	Fluvaquentic Dystrudept	Loam	Floodplain	Recent loamy deposits from local[††] sources
Kenefick	Ultic Hapludalf	Sandy loam	Pleistocene terraces	Older sediments on former floodplains
Kian	Aeric Endoaquept	Sand	ACWBs on tributary floodplains	Holocene channel fills
Mantachie	Fluvaquentic Endoaquept	Sand, clay	Floodplain	Dominantly local[††] alluvium
Ozlas	Aeric Dystraquert	Clay[*]	Floodplain	Dominantly upper Neches alluvium
Pophers	Fluvaquentic Endoaquept	Loam	Seasonally flooded floodplains	Recent alluvium overlying older alluvial soils
Pluck	Fluvaquentic Endoaquept	Sand	Abandoned tributary channels	Holocene channel fills
Simelake	Aeric Dystraquert	Clay[*]	Clayey Holocene alluvium; floodplain flats	Dominantly upper Neches alluvium
Turkey	Typic Quartzimpsamment	Sand	Sandy Pleistocene terraces	Older sandy alluvium with minimal pedogenic development

Continued

Table 1.2. Continued

Series	Taxonomy	Texture	Setting	Interpretation
Urbo	Vertic Epi- aquept	Clay*	Floodplain	Mixture of dom- inantly upper Neches al- luvium and local sediment sources
Votaw	Oxyaquic Quartzip- samment	Sand	Pleistocene ter- races influenced by seasonal high water table	Older sandy alluvium with minimal pedogenic development

SOURCE: US Department of Agriculture soil databases. Taxonomy is according to the US system.

Soil taxonomic names are often bewildering to non-pedologists and are included here for the benefit of (fellow) soil nerds. You do not need to understand these terms to get the major points here.

*Indicates smectitic clays with vertic properties.

†Sediment likely transported from portions of the basin upstream of the study area, where smectitic clays are more common.

††Sediment mainly derived from tributaries and lower drainage basin.

The upper Neches River drains an area of relatively high cation exchange capacity soils, including Vertisols of the Blackland Prairie region. Tributaries in the lower coastal plain, however, do not drain areas of Vertisols and do drain extensive areas of ultisols dominated by low-activity clays. Clay mineralogy is therefore a rough, first-order indicator of a sediment source dominated by the river vs. tributaries. Soil geography there- fore reveals "geomorphological filtering" analogous to ecological filtering.

Soil drainage is linked to the hydrological variations described above. Key factors are relative elevations, distance from channels, and position on convex, flat, or depressional topography. Time and age also play a role—providing time for pedogenesis to produce features such as soil structural aggregates and argillic horizons (clay-rich subsoils).

Ecological Sites

Ecological mapping at a scale comparable to soil mapping is recent for the lower Neches region, and some of the types identified in the US Department of Agriculture's Ecological Site Descriptions (ESD) are still provisional. Ecological sites are recognized and described with respect to characteristics that differentiate them from other sites in their ability to produce and support a characteristic plant community (i.e., selection via ecological filtering). An ESD is defined as a distinctive kind of land based on recurring soil, landform, geological, and climate characteristics that produces distinctive kinds and amounts of vegetation.

ESDs typically have multiple soil series associated with them rather than a one-to-one correspondence with soil types. However, none of the soil series listed in Table 1.2 is associated with more than one ESD. The ESDs identified so far do give an idea of the transformative history and future possibilities. Each has multiple potential states and plant communities identified in a state-and-transition model for each ESD. Natural and semi-natural factors such as fire regimes, flooding frequency, vegetation invasions, and succession may drive transitions, as can deliberate actions of humans too. Each ESD has a distinctive set of dominant plant species. In the study area, there exist eleven recognized ecological sites, with two to seven potential plant communities each (forty-three total).

Recursivity: Feedbacks and Mutual Adjustments

Fluvial geomorphic processes, driven in many cases primarily or entirely by internal interactions within the fluvial system, create, modify, and change landforms, topographic settings and substrates that are characterized by distinct hydrological states, and characteristic suites of soil types. The hydrology, soils, and landforms create distinctive ecological sites with characteristic vegetation communities. I've told the story this way for simplicity's sake, and it is not wrong. But it does imply mainly one-way cause-effect links, while in reality constant reciprocal interactions and mutual adjustments occur.

As is usually the case, plant-soil feedbacks are common. The oaks and pines (*Quercus* spp., *Pinus* spp.) that dominate the over-story on several ecological sites, for example, acidify soil with their litter, creating advantageous conditions for themselves and other acid-tolerant plants. Forest cover may also be necessary to formation of the argillic horizons found in the more strongly developed soils (discussed more in Chapter 2). Those clayey subsoils themselves, as well as root systems, are key determinants of soil moisture flux and storage. For another example, Cherokee sedge (*Carex cherokeensis*) is common in the region and dominates the understory at some sites. It has a distinctive bunched or clumped growth habit, a physiognomy that fosters local spatial patchiness of soil moisture, nutrients, and organic matter.

Vegetation also influences avulsions. Logjams can deflect or displace flow from channels to trigger crevasses. On the levees, uprooting of larger trees leaves local topographic gaps and low-resistance locations that can lead to crevasses. Feedbacks between logjams and avulsions may also occur, as avulsions may split a larger channel into two smaller ones. The narrower channels may not be wide enough to allow larger logs to be transported downstream, initiating logjams. Vegetation also plays a key role in stabilizing point and lateral bars as they transition from fluvial bedforms to floodplain landforms.

Sediment and soil texture also feature important feedbacks. Fluvial processes such as avulsions and cutoffs create a variety of landforms and depositional environments, which in turn are often associated with either dominantly sandy or clayey deposits (as evident in Table 1.2). Sand has low cohesion and is more highly erodible, facilitating, for example, lateral channel migration where exposed on channel banks and potential re-excavation and reoccupation of previously abandoned channels. Cohesive clays have the opposite effect.

These examples of feedbacks among geomorphic processes, landforms, hydrology, soils, and biota as the lower Neches River landscape comprise an incomplete but illustrative selection that could probably be expanded indefinitely.

Down in those Neches River bottomlands, landscape evolution is transformative in many ways, and reciprocal interactions abound. The landscape and its components emerge nondeterministically. While the story is most easily told as a geomorphologically driven tale, the landscape (continuing into the present) is a shifting mosaic of habitats, soils, and landforms reflecting the transformative reciprocal emergent evolution.

————————

This was a long chapter but is by far the longest you'll face. We have explained the whole *Mysterious Ways* mission, laid out some important conceptual frameworks, let you know where I'm coming from, and given some examples. The latter are important—and I believe, necessary—because in this type of science you must always, always connect ideas and theories with real landscapes.

Okay, then, let's move on.

NOTES

1.   H. Gao, F. Fenicia, H. H. G. Savenjie, "Are Soils Overrated in Hydrology?" *Hydrology and Earth System Sciences* 27 (2023): 2607–2620.

2. Regolith refers to all weathered, unconsolidated, or poorly consolidated material overlying unaltered bedrock or sedimentary deposits. Regolith includes soil and highly weathered rock (sometimes referred to as saprolite, saprolith, and saprock).

3. See T. M. Brown and M. J. Kraus, "Time-Stratigraphic Reconstruction and Integration of Paleopedologic, Sedimentologic, and Biotic Events (Willwood Formation, Lower Eocene, Northwest Wyoming, U.S.A.)," *Palaios* 8 (1993): 68–80; G. J. Retallack, "Late Oligocene Bunch Grassland and Early Miocene Sod Grassland Paleosols from Central Oregon, USA," *Palaeogeography, Palaeoclimatology, Palaeoecology* 207 (2004): 203–237; G. J. Retallack, "Cenozoic Paleoclimate on Land in North America," *Journal of Geology* 115 (2007): 271–294; Y. Bouchenak-Khelladi, E. C. February, G. A. Verboom, F. C. Boucher, "C4 Grass Functional Traits Are Correlated with Biotic and Abiotic Gradients in an African Savanna," *Plant Ecology* 221 (2020): 241–254; R. Rye and H. D. Holland, "Paleosols and the Evolution of Atmospheric Oxygen: A Critical Review," *American Journal of Science* 298 (1998): 621–672.

4. "Gaian" refers to the Gaia theory of James Lovelock, which holds that the biosphere modifies the global environment, particularly the chemical composition of the atmosphere, to maintain conditions favorable to life and that the biosphere and abiotic environment have coevolved. See J. Lovelock, *The Ages of Gaia: A Biography of Our Living Earth* (New York: Norton, 1995). A. G. Lapenis, "Directed Evolution of the Biosphere: Biogeochemical Selection or Gaia?" *Professional Geographer* 54 (2002): 379–391; J. D. Phillips, "Emergence and Pseudo-Equilibrium in Geomorphology, *Geomorphology* 132 (2011): 319–326.

5. *Karst* refers to terrain with distinctive hydrology and landforms that arise from a combination of high rock solubility, conditions humid enough to enable dissolution, and well-developed secondary (fracture) porosity. Such areas are characterized by sinking streams, caves, enclosed depressions, fluted rock outcrops, and large springs and most often occur in limestone rock.

6. The Cretaceous is a geological period (the third, final, and longest of the Mesozoic era) that lasted from about 145 to 66 Ma. The Eocene epoch lasted from 56 to 33.9 Ma, the second epoch of the Paleogene period of the Cenozoic era. The Miocene is the first geological epoch of the Neogene period, lasting from about 23 to 5.3 Ma.

7. This information comes primarily from M. M. El Aref, A. Salama, and M. Hammed, "Morphotectonic Evolution of Qaret El Sheikh Abdallah Depressions and Denuded Paleokarst in the White Desert, El Bahariya-Farafra Karst Territory, Egypt," *Egyptian Journal of Geology* 65 (2021): 27–53; N. S. Embabi, *Landscapes and Landforms of Egypt: Landforms and Evolution*, World Geomorphological Landscapes Series (Berlin: Springer, 2018).

8. This highly abbreviated inventory of (paleo)karst features is from E. A. Khalaf, "Karst Heritage as a Tourist Attraction: A Case Study in the White Desert National Park, Western Desert, Egypt," *Geoheritage* 14 (2022): 94.

9. The Holocene (modern) epoch began about 10 to 12 Ka. Reference: K. Kinderman et al., "Palaeoenvironment and Holocene Land Use of Djara, Western Desert of Egypt," *Quaternary Science Reviews* 25 (2006): 1619–1637.

10. Recursivity in karst: I. Bárány-Kevei, "Geoecological System of Karst," *Acta Carsologica* 27 (1998): 13–25.; O. Bonacci, T. Pipan, and D. C. Culver, "A Framework for Karst Ecohydrology," *Environmental Geology* 56 (2009): 891–900. J. D. Phillips, "Biogeomorphology and Contingent Ecosystem Engineering in Karst Landscapes," *Progress in Physical Geography* 40 (2016): 503–526. Oases, groundwater, and role of paleokarst depressions: A. M. Sharaky, S. I. Gaber, K. A. E. Gelil, and S. H. Abdoun, "Groundwater Potential in the Bahariya Oasis, Western Desert, Egypt," in *Groundwater in Egypt's Deserts*, ed. A. Negm and A. Elkhouly (Berlin: Springer, 2021) 201–242; M. M. El Aref et al, endnote 7; Concentration of resources, biota, and human activity in oases: E. A. Khalaf, endnote 8; K. Kinderman et al., Palaeoenvironment and Holocene Land Use of Djara, Western Desert of Egypt," *Quaternary Science Reviews* 25 (2006): 1619–1637.

11. J. Cavelier et al., "The Savannization of Moist Forests in the Sierra Nevada de Santa Marta, Colombia," *Journal of Biogeography* 25 (1998): 901–912.

12. Evidence for occurrence of savannas, forest-to-savanna, and savanna-to-forest transitions prior to human impacts comes from: T. van der Hammen and M. L. Absy, "Amazonia during the Last Glacial," *Paleogeography, Paleoclimatology, Paleoecology* 109 (1994): 247–261; R. Marchant et al., "Colombian Vegetation at the Last Glacial Maximum: A Comparison of Model- and Pollen-Based Reconstructions," *Journal of Quaternary Science* 19 (2004): 721–732; B. M. Flores et al., "Soil Erosion as a Resilience Drain in Disturbed Tropical Forests," *Plant and Soil* 450 (2020): 1–25.

13. This section is largely based on J. D. Phillips, "Universal and Local Controls of Avulsions in Southeast Texas Rivers," *Geomorphology* 13 (2011): 17–28; J. D. Phillips, "Anastamosing Channels in the Lower Neches River Valley, Texas," *Earth Surface Processes and Landforms* 391 (2014): 888–1899; J. D. Phillips, "Tipping Points in Texas Rivers," *Earth Surface Processes and Landforms* 43 (2018): 1768–1781. I included a more detailed analysis of the lower Sabine River (Texas and Louisiana) in *Landscape Evolution*. Rather than summarize that here, I decided to analyze the lower Neches in a similar context. Though the rivers have surprisingly different avulsion regimes (as outlined in the references above), a similar story of TREE emerged.

14. The Quaternary period, characterized by alternating "ice ages" (colder climates, advancing glaciers, and lower sea levels) and warmer interglacial

periods, began 2.6 Ma. The Pleistocene epoch, accounting for most of the Quaternary, ended about 12 Ka.

15.   J. D. Phillips, "Hydrological Connectivity of Abandoned Channel Water Bodies on a Coastal Plain River," *River Research and Applications* 29 (2013): 149–160.

16.   Containing the clay mineral smectite. Smectitic clays swell when wet and shrink when dry.

2

Supra-Organic Landscapes

SUPRA- VS. SUPER-

The distinction between biotic/abiotic is real and useful. Certain aspects of Earth's landscapes are best studied from a purely, or mostly, biological or geophysical perspective. However, for Earth surface systems (ESS), landscapes, and the planet as a whole, the distinction can be a bit misleading. For instance, atmospheric sciences mostly fall within the realm of abiotic physics and chemistry. However, the chemical composition of the atmosphere is created and maintained by the biosphere, and the atmosphere and biosphere coevolved. Similarly, many aspects of meandering rivers are effectively studied based entirely on geophysical principles and tools. However, the paleoenvironmental record shows that (except in the case of some very cohesive materials) meandering rivers did not exist until biological evolution produced vascular plants, and studies show that in the vast majority of cases a meandering pattern cannot be formed or maintained in the absence of vegetation. And in the Devonian (about 360 to 420 million years ago), the adaptation of life to dry land triggered a revolution that encompassed landforms, soils, and climates.

Now that life is here on the planet since at least the Paleozoic, any variations or changes (locally or globally) in atmospheric and ocean chemistry and circulation, volcanic and tectonic activity, solar inputs, and cryosphere (ice) dynamics have significant

Mysterious Ways. Jonathan D. Phillips, Oxford University Press. © Oxford University Press (2025).
DOI: 10.1093/9780197755129.003.0002

to overwhelming influences on habitat, stresses, resource availability, and dispersal for organisms. And now that life is here
on the planet, no aspect of the Earth surface or near-surface
environment is unaffected by biota. So what if, as organisms cannot exist independently of their environment, ecosystems are the
most important unit of selection and evolution, as some ecologists
have suggested? What if, as some hydrologists and geomorphologists have suggested, forms and fluxes at Earth's surface
do not occur independently of biota? And what if the soil, encompassing intertwined, coevolving biotic and abiotic aspects by
definition, is representative in this respect of landscapes and
ESS as a whole? This brings us to the supra-organic concept of
landscapes.

Given the highly interconnected, mutually adjusting, coevolving
nature of landscapes, it is tempting to start to think of them as a
single complex machine—or even as a single organism or super-
organism. Indeed, Clements explicitly described plant communities
in this way. In his early versions of the Gaia theory (then the Gaia
hypothesis), James Lovelock used the super-organism metaphor
for the entire planet. I am *not* going to argue that landscapes are,
even metaphorically, machines or super-organisms. I *am* going to
make a case for landscapes as *supra*-organisms.

First, let's explore the difference *ra* vs. *er* makes. *Super*-
organism refers to entities larger than an individual, including
multiple organisms, that act or function as a single being. Scientists have mostly used super-organic concepts metaphorically, to
highlight the interconnectivity of phenomena ranging from social
insects to Earth's biosphere. Human societies, cities, and cultures
have also been studied and described by social scientists using
super-organic metaphors. Describing ecological systems and the
biosphere as super-organisms is inaccurate in equating phenomena applicable to individuals to those of larger systems, misleads
non-scientists, and implicitly separates biota from the abiotic environment (among other things). *Supra*-organism has sometimes
been used interchangeably with super-organism—sorry for any

confusion there, but I can barely control my own semantics, much less anyone else's, and I could not come up with a better term for what I'm getting at here. Supra-organism was coined in part to get at the tightly interconnected, coevolving, and biotic and abiotic aspects of ecosystems and other ESS, *without* the idea that they are single purposeful beings or other baggage of super-organic concepts. For instance, Lovelock's Gaia theory, introduced as the Gaia hypothesis in the 1970s, was criticized for claiming that Earth is a super-organism. As Lovelock was using the super-organism term metaphorically rather than literally (as his critics charged), by 1995 he had switched to the term *supra-organism*. In 1995, the mathematical ecologist Robert May, separating the key scientific ideas from some non-scientific notions that Gaia theory inspired among laypersons, declared, "We do not have to embrace the wilder poetic flights of the Gaians to acknowledge that ecosystems can usefully be regarded as supra-organisms for many discussions of the way biological and physical processes entwine to maintain the biosphere." He gave examples of the role of plants in cloud formation and structure (via hydrological transformations) and of the water and carbon cycles. These are best understood, May argued, as functional aggregates. Ditto for soil formation, where the effective units are not individual organisms but rather a diverse array of functional communities of organisms. Organisms cannot exist without abiotic environments, May said (a common theme in discussion of supra-organic concepts), and the supra-organism includes both.

Canadian ecologist Stan Rowe, a leading proponent of supra-organic ideas, in 2001 directly contrasted supra- and super-organisms: "Earth should not be confused with organisms. Earth is not a super-organism. Earth is supra-organic." Similarly, in 2010 agricultural reformer and plant geneticist Wes Jackson also explicitly differentiated supra- from super-: "All organisms . . . are embedded within a living ecosphere, a supra-organism, not a super-organism." Jerry Brunetti (2014), whose mission is treating farms as ecosystems, promoted a view of soil as a supra-organism.[1]

I define a supra-organism as an Earth surface system that includes multiple interacting biota and abiotic factors and that changes over time and responds to disturbances and changes as a single unit (Figure 2.1). This does not at all preclude responses or changes by specific components of the system, but it does imply that these are linked to system-level responses. A supra-organism is an environmental system whereby neither abiotic factors nor biota operate or develop independently of each other. This means that, in addition to evolution of its constituent elements, supra-organisms evolve at the system or supra-organic level. Therefore, my task here is to show that landscapes respond and adapt to environmental change as functional units, independently of responses and adaptations of their biotic and abiotic constituents, and that they evolve as a unit, exhibiting directional change steered by selection.

SUBDISCIPLINARY SUPRA-ORGANICISM

The twelve-syllable subheading refers to trends toward at least an implicit supra-organic view reflected in the recent emergence of several hybrid subdisciplines in the landscape sciences. In Chapter 1, we saw that for a long time studies of bio-geo-hydro-soil interactions were predominantly one-way in terms of causality, as opposed to reciprocal interactions. The birth of new subdisciplines reflects the emergence of recognition that organisms, soils, landforms, hydrology, and climate must be understood as operating, responding, and evolving together.

The oldest of these new (sub)disciplines is *biogeochemistry*, dating back more than a century to Vernadsky in the 1920s. This melding of biology, geology, and chemistry is concerned with flows, cycles, and transformations of elements such as carbon, oxygen, nitrogen, and phosphorus. These involve both bio- and geochemical processes, geophysical and biological transport and exchanges, and interactions of the biosphere, atmosphere, and lithosphere.

Figure 2.1. A painting by Czech artist Petr Mores to show the interrelationships of *Picea abies* (Norway spruce trees, both dead and alive), landforms, and geomorphic processes as represented by topography, soils, and underlying geology. In high altitudes of the Sumava Mountains of the Czech Republic, these elements, along with hydrology and other biota, coevolve and respond as a unit to environmental changes and disturbances.

J. D. Phillips and P. Šamonil, "Biogeomorphological Domination of Forest Landscapes: An Example from the Šumava Mountains, Czech Republic," *Geomorphology* 383 (2021): 107698.

Though it never spawned its own dedicated journals or organizations, there also exists a tradition of *geoecology*, focusing
mainly on interrelated geomorphological and ecological processes.
More recently *ecohydrology* has come into focus, emphasizing
interactions and feedbacks between ecological and hydrological
systems. *Biogeomorphology* (or ecogeomorphology) deals with reciprocal interactions between landforms and surface processes on
one hand and the functioning of ecological systems and organisms on the other. *Hydropedology* was proposed to link traditional
pedology with soil physics and hydrology. *Geobiology* explores the
relationship between life and the Earth's physical and chemical
environment.

Ecosystem, critical zone, Earth system, and climate science
have also become recognizable subdisciplines. The name *ecosystem science* emphasizes that ecosystem studies are not just ecology but also have strong links to biogeochemistry, aquatic ecology,
soil science, hydrology, ecological economics, and conservation
biology, as well as landscape ecology and global ecology. Analo-
gously, the term *climate science* represents an effort to expand the
scope and the perception of climate research. *Critical zone science*
is an integrated approach to the study of rock, regolith, soil, water,
biota, and atmosphere interactions near Earth's surface, with the
critical zone defined as the planet's permeable near-surface layer
from groundwater to treetops. The need to understand the many
highly interconnected components and reciprocal interactions of
the atmo-, hydro-, litho-, and biospheres, particularly at very broad
(often global) scales, gave rise to *Earth system science*.[2]

No aspect of Earth's landscapes or systems can be fully understood in isolation from the others, and ESS are typified by constant
internal and external feedbacks and reciprocal interactions. The
advent of the subdisciplines above and their associated terminology and concepts recognizes these truths, and while it does not
necessarily imply full embrace of landscapes as supra-organisms,
taken together these developments are wholly consistent with
supra-organic ideas.

Selection and evolution at the ecosystem level is discussed later in this chapter.

EXTENDED COMPOSITE PHENOTYPES

The phenotype is the genetic expression evident in an organism's body (as opposed to the genotype). British biologist Richard Dawkins in 1982 developed the concept of the extended phenotype, which encompasses all the effects of genes on the environment, expressed via the actions or behavior of an organism. Common examples are ant and termite mounds and beaver dams, which have especially strong impacts on soils, landforms, and hydrology. Dawkins argued that the alleles supporting the environmental modifications—which facilitate survival and reproduction of the responsible organisms—are selected for. Organisms and their biotopes can define and construct each other, Dawkins held, including an explicitly reciprocal aspect.

It occurred to me in my field studies that in many cases soils and landforms reflect the combined, contemporaneous, environmental effects of multiple organisms. I therefore proposed in 2009 that soils, and in 2016 that landforms, are extended *composite* phenotypes (ECPs), a perspective that helped lead to my supra-organic view. Dawkins has not commented on the ECP idea, but there is a good chance he would not be fully supportive, as he favors a strict definition of extended phenotypes based on a biotope's role in increasing the representation of sources of genetic variation (i.e., specific alleles). An ant mound would not qualify as an extended phenotype only because it improves habitat and increases the likelihood of survival and reproduction of the ants—you must also show that superior mound-building genes are preferentially selected for. Others apply a broader definition, requiring only that the extended phenotype supports persistence and growth of specific species. Still others prefer the niche construction concept, whereby organisms modify the environment to create niches that, in turn, exert selection pressure. Niche construction requires only ecological filtering rather than selection at the gene level, but it does not always lead to features such as soils or landforms that are expressions of specific (groups of) organisms. Rather than add additional terminology, I simply acknowledge a broad version of the extended phenotype and do not engage in the debate centering on the role of alleles, genes, and individuals as agents of biological selection or add additional terminology (because it makes no difference with respect to the ECP concept).

A few soil scientists have embraced the ECP idea, and more evidence has been found demonstrating that some of the mechanisms involved in pedogenesis exert selective effects on organisms that in turn influence pedogenesis. Several other pedologists anticipated the idea without using the ECP terminology. It has long been recognized that soils may represent biotopes of multiple organisms. Some grassland soils (Mollisols or Chernozems in the major soil classification systems) are phenotypes of both grasses and earthworms, for instance, and soil biomantles (see below) may be extended phenotypes of multiple organisms (Figure 2.2). He did not address phenotypes, but Dutch soil scientist Nico van Breemen in 1993 described soils as "biotic constructs" that are affected by biota to increase net primary productivity. Van Breemen specifically listed effects of organisms on soil ecology, chemistry, texture, structure, erosion, and organic matter content that make Earth's surface more conducive to plant growth. Some pedological effects of organisms also give particular species advantages by inhibiting or stressing competitors, he noted.

The upper layers of soil are often continuously mixed by biota— tunneling, burrowing, foraging, root growth and decay, uprooting, etc.—and this is reflected in the *biomantle* concept, most thoroughly discussed by Don Johnson (1934–2013). The roots of this idea can be traced to American geologist Nathaniel Shaler and Charles Darwin in the late nineteenth century, but conceptual (and more recently simulation) models incorporating continuous

Figure 2.2. Semi-arid soil from South Australia. In addition to other formative factors, this soil is an extended composite phenotype of the grasses and other plants and various soil invertebrates.

bioturbation have become widely accepted only in the last two decades. Vigorous bioturbation is more common than you might think and has even been found in Antarctica!

Australian soil and plant scientists William Verboom and John Pate independently developed the *phytotarium* concept in the 2000s, based mainly on work in eucalypt forests in Australia. In phytotaria, plants, along with their associated microbial communities, undertake niche construction at specific vertical layers to maximize access to water and nutrients. They presented empirical evidence showing that in some cases vertical texture contrast regoliths may be biogenic, and not just due to bioturbation and indirect effects already known to influence vertical texture contrasts. Some trees have a role in synthesizing specific clays and can directly redistribute clays within soil profiles. This makes a strong argument that some texture-contrast profiles are extended phenotypes of tree species, though Pate and Verboom do not use

ECP terminology. They also showed strong vegetation roles in creating "clay pavements" in some western Australian regoliths.[3]

With respect to landforms, some (e.g., volcanic features, hyperarid desert dunes, glacial scours) are independent of direct biological influences and could be considered abiotic. However, independently of the soils they may host, many landforms are ECP. Geomorphic processes and landforms are often strongly influenced by biota, ecological and geomorphological processes are frequently closely intertwined, and landforms and surface processes are key components of habitats for many organisms. Beyond this, in recent years it has also become widely recognized that tight coupling and reciprocal interactions transcending mutual influences are common, such that landforms and biota (from individuals to communities) develop interdependently. Dov Corenblit in France and Woody Cotterill in South Africa have been particularly important in exploring not only reciprocal interactions but also relationships with ecosystem engineering, niche construction, and coevolution—that is, more explicitly linking concepts of geomorphology, ecology, geological evolution, and biological evolution. Daehyun Kim and Keonhak Lee of Seoul National University argued that the role of multiple organisms in shaping landforms should be incorporated into a revised ecosystem engineering concept and gave empirical examples in terrestrial and coastal dune ecosystems, crab–plant interactions on tidal wetlands, pocket gopher–plant relationships in grasslands, and biological invasions by exotic mollusks and macrophytes.[4]

There is a four-part case for landforms as ECP: process-form relationships, synchrony, selective pressure, and positive feedback. Process-form relationships demonstrate that actions of organisms can create landforms or strongly influence their development and characteristics. Synchrony implies that landforms and biota developed contemporaneously—a landform cannot be a phenotype if it appears in the paleoenvironmental record *before* the organism(s). The criterion of selective pressure holds that the landform creates or facilitates environmental conditions that differentially advantage (or disadvantage) specific organisms. The positive feedback criterion is that the selective pressure should provide advantages to the organism(s) associated with creation of the landform. Examples of features meeting all four include biogenic landforms such as coral reefs, stromatolites, tree uproot pit-mounds, peat bogs, and termitaria. Other examples include many karst landforms, biogeomorphic interactions in vegetated sand dunes and coastal wetlands, the coevolution of vascular plants and meandering rivers, and examples of landscape metamorphosis driven by biogeomorphic interactions (e.g., when wholesale change is driven by introduction of a new species).[5]

So how does this relate to the broader argument? Because many landforms and soils are ECP, as the biota within them respond to environmental change, the soils and landforms (and hydrologic and other biogeochemical fluxes and cycles) that are phenotypes of the biota also respond. This coherent response supports the idea of landscapes as supra-organic.

STATE FACTOR MODEL

To recap: the supra-organism concept is *not* meant to imply an equivalency between landscapes and organisms (or a landscape super-organism). It does imply that landscapes function, respond to change, and evolve or develop as a single, integrated unit (again, this does not preclude responses and evolution of landscape subsystems). Supra-organic also does not infer any goal functions, as systematic developmental trends and even self-organization can be explained as emergent outcomes (we'll delve into this in Chapter 7). The emergence results from selection principles dictating that more durable and/or efficient patterns, structures, and morphologies have a greater probability of occurrence and persistence than less-durable and less-efficient ones.

What would constitute convincing evidence for ESS as supra-organisms? First, supra-organic landscapes should have multiple interrelationships among and within abiotic and biotic components. They should exhibit mutual adjustments and feedbacks among geology, hydrology, landforms, soils, biota, and climate (at least at the microclimate scale, but sometimes more broadly). This is well established, if not axiomatic, at this point. Second, significant changes to any part of a supra-organic landscape should lead to broader, system-level changes; responses and adaptations would be manifested at the landscape level and involve multiple system components. Third, evolutionary or developmental changes should occur contemporaneously within a supra-organic landscape system—not always or necessarily at the same time and certainly not at the same rate, but at least with temporal overlaps or a close sequence of changes.

The criteria above are not embedded within existing super-organism (or not) debates. For instance, both the Gleason/community assembly concepts or Clementsian ecological succession could potentially meet (or not) the criteria, depending on the existence, and rates, of feedback between organisms and abiotic factors. Further, supra-organic landscapes need not be dynamically stable—in fact, ESS meeting the criteria are quite likely to be complex and prone to instability.

In the rest of this section, we explore the supra-organic nature of landscapes in the context of a classic concept of ESS: the state factor model.

BACKGROUND

Back in 1883, Dokuchaev began revolutionizing the study of soils (and landscapes more generally), showing that soils are not just geological debris with some plants and organic matter on top. Rather, soils are distinct objects produced by the combined (and interacting) effects of geology, climate, topography, plants, and animals. Hydrology was implicitly included through climate (via moisture supply) and topography, via controls on flows of water in the landscape. Dokuchaev's was the first explicit factorial model, where soils and soil properties were seen as the function of soil-forming factors (state factors). In the anglophone world, the most influential of several early-twentieth-century elaborations of the state factor model was Hans Jenny's (1899–1992) 1941 book *The Factors of Soil Formation*. From this comes that staple of intro-ductory soil science, physical geography, and ecology classes, the "*clorpt* equation":

$$S = f(cl, o, r, p, t) \ldots.$$

S is the soil (or a particular soil property), and the others represent, respectively, climate, organisms (biosphere), relief (topography and drainage), parent material (geology), and time. The ellipsis (trailing dots) recognizes the soil-forming factors that are important locally but not in all landscapes, such as rates of sea-level change in coastal settings.

The genesis of the state factor model was in soil science, but it was quickly adapted to (or sometimes semi-independently devel-oped) in many explanatory frameworks rooted in geographical and ecological thinking. The state factor model was explicitly applied in plant ecology and ecosystem science in the mid- and late twentieth century. The most detailed factorial model to date was devel-oped by Greg Pope (Montclair State University) and colleagues in 1995 to explain geographical variations in rock weathering. Richard Huggett of Manchester University in the 1990s developed a state factor model for ESS in general. Factorial models have also been applied in other aspects of geomorphology and in Earth system sci-ence. In paleoecology and paleopedology, the state factor model is the main (though often implicit) underpinning of the fundamen-tal logic—given that ecological systems and soils are functions of environmental factors, historical evidence is interpreted in terms of what it tells us of those environmental (state) factors.[6]

Though Jenny hinted at a dynamical systems-type formulation of the "clorpt" model, early- and mid-twentieth-century factorial concepts were generally not explicitly framed in those terms. Dokuchaev and Jenny both acknowledged that state factors and soils themselves affect and are affected by each other. In the 1990s, the state factor model was expressly reconfigured in a nonlinear dynamical systems framework, explicitly incorporating feedbacks and mutual adjustments. Thus the "clorpt" equation (or its equivalent) becomes the clorpt equations:

$$S = f(cl, o, r, p, S) \ldots.$$
$$cl = f(cl, o, r, p, S) \ldots.$$
$$o = f(cl, r, p, S) \ldots.$$
$$r = f(cl, o, p, S) \ldots.$$
$$p = f(cl, o, r,, S) \ldots.$$

Note two critical things: First, the t for time is missing. That's not because time is ignored but because this formulation is a dynamical systems formulation more properly rendered as differential equations reflecting change over time (I've spared you that more complicated and cluttered notation). Second, the "dot factors" will be quite different—for instance, climate depends on aspects of solar radiation and general atmospheric circulation that are not directly relevant to the other state factors.

LANDSCAPE STATE FACTOR MODEL

In a 2019 article, I proposed a general interactive state factor model for landscapes as shown below, implying that the state of a landscape is a function of the combined effects of substrate, propagules, climate, biotic establishment, hydrogeomorphology, and soil, interacting as shown in Figure 2.3. *Substrate* corresponds to the geology or parent material factor included in most factorial models. *Climate* specifically indicates climate attributes such as insolation (solar radiation input), moisture and temperature regimes, and other climate variables that affect biological habitat and soil formation. Biota are represented by two factors here. *Biotic establishment* denotes the colonization and growth of biota within the landscape. *Propagules* represents the availability of potentially reproductive individuals or the supply of seeds, root-stock, etc. to colonize a landscape. This factor is linked to the biota within the landscape and to external seed sources, dispersal mechanisms, and metapopulations. *Hydrogeomorphic context* includes hydrologic and geomorphic processes and setting, including the topographic and drainage elements incorporated in topographic or relief state factors as well as hydrogeomorphic mass fluxes into and out of the landscape. Hydrogeomorphic context also includes disturbances such as slope failures, floods, and cyclones. *Soil* has been significantly modified from its parent material by chemical, physical, and biological processes, and thus differs from substrate. A landscape in early stages of development (or re-development following disturbance) may have little or no soil, only substrate. Likewise, in some landscapes that have developed thick soils, the soil factor will subsume the substrate factor. In analyzing the model, I tested three versions: with soil and substrate, and with either one or the other.

Graph theory is a set of mathematical techniques for analyz-ing and describing the properties and dynamics of networks—sets of elements or components (the *nodes* of the graph) and their interconnections (the *links* or *edges* of the graph). Some as-pects of so-called algebraic or spectral graph theory are closely linked to nonlinear dynamical systems theory. Graph theory there-fore includes tools very well suited to representing and analyzing landscapes and ESS.

The state factors in Figure 2.3 are graph nodes, with links or edges between them based on causal connections. Figure 2.3 doesn't propose anything new with respect to individual causal relationships. Soil is a function of substrate (parent material), cli-mate, biota, and hydrogeomorphic context, as in the traditional state factor model. Establishment of biota is controlled by the available propagules and habitat suitability, determined by climate,

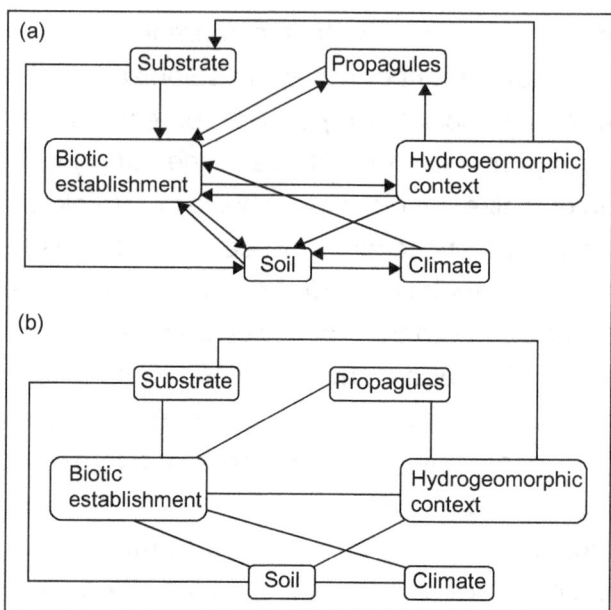

Figure 2.3. State factor model of ecosystems as an unweighted graph; directed (top) and undirected (bottom). Reproduced with permission from J. D. Phillips, "State Factor Analysis of Ecosystem Response to Climate Change," Ecological Complexity 40 (2019): 100789.

substrate, hydrogeomorphic context, and soil properties. Soil impacts climate, mainly via soil moisture dynamics and albedo (reflectivity of solar radiation), and biota affect geomorphic and hydrological processes. Hydrogeomorphic fluxes are often important in the dispersal of propagules (e.g., seeds or spores borne by wind or water), and once organisms are established, the biota themselves become an important influence on propagule supply. Figure 2.3b is an undirected graph, such that lines indicate which components are interconnected. Links in a graph can also be directional and may be numerically weighted. An undirected, unweighted graph is the simplest and (importantly) most generally applicable form.

Here is where, in a scientific publication, I would plunge into several pages worth of mathematical, systems, and graph theory detail. These are important, but they are not of interest to the non-specialist reader (and difficult for those not trained in graph theory). So, for those who are interested in those details, or skeptical of my claims below, I refer you to a footnote and proceed to the bottom line.[7]

Properties of the landscape state factor graph model reflect its complexity or simplicity in terms of sensitivity to changes in any component(s) to the system as a whole. These properties also indicate something called graph energy (a mathematical, not a physics term in this case), which indicates the extent to which changes to or responses of the graph/network/landscape tend to reverberate through the system. Yet another metric indicates graph synchronization, the extent to which changes occur contemporaneously or in a regular sequential order vs. at different times or in variable sequences throughout the network. Synchronization can be literal, in terms of time dynamics, or inferential, in terms of the extent to whether effects of changes in one part of the graph can be predicted other parts. The analysis indicates that landscape systems are sensitive to changes in any component, that responses reverberate through the system, and that there is a high level of synchrony. The results in a qualitative sense are identical in all three versions (substrate only, soil only, soil and substrate), though the numbers, as you would expect, come out differently.

Recall that above it states that in addition to multiple interrelationships among and within abiotic and biotic components, and mutual adjustments and feedbacks, if landscapes are supra-organisms they should have two other key characteristics: responses and adaptations are manifested at the landscape level and involve multiple system components, and evolutionary or developmental changes should occur contemporaneously within a supra-organic landscape system. Results of the graph theory analysis indicate that this is indeed the case.

In landscape sciences, purely abstract models are just parlor games without some ground truth to back them up. The results of the landscape state factor analysis also yielded some mathematical criteria for judging graph models of specific landscapes with respect to supra-organic behavior. I applied these to three different situations: forested hillslopes, barrier island sand dunes, and wetlands of the Florida Everglades. All three passed the supra-organism test, indicating that landscapes respond and adapt to environmental change as functional units, independently of their biotic and abiotic constituents.[8]

ECOSYSTEM EVOLUTION

The supra-organic notion of landscapes and ESS ties into a traditional debate in ecosystem science, ecology, and evolutionary biology: Do ecosystems evolve? Specifically, can ecosystems evolve as an entity, at the ecosystem level? The supra-organic concept says yes. Depending on the scale at which either landscapes or ecosystems are defined, in some cases they are essentially— or even exactly—commensurate. So, if landscapes are supra-organisms, at least some are ecosystems, and they can evolve as such. But this has been pondered long before me.

The idea that ecological systems can evolve at the system level due to selection processes predates the explicitly supra-organic concepts already mentioned. Charles Smith's overlooked biotic/abiotic conceptual model, first proposed in 1986, holds that biological adaptations are the structural attributes that mediate energy dissipation. Therefore, Smith, of Western Kentucky University, argues that it is the properties of spatial interaction that evolve, not the organisms themselves, and both responses to and influences on the abiotic environment are guided by selection for more efficient means of turning over material resources. In 2002, Andrei Lapenis coined the term *biogeochemical selection* and suggested selection that favors faster energy and nutrient (re)cycling can account for "Gaian" evolution at the scale of ecosystems and the biosphere, without invoking super-organicism. Lapenis also traced those ideas to evolutionary, ecological, and biogeochemical theories that arose in Russia in the early twentieth century from Vernadsky, soil geographer R. V. Rizpolozhensky (1847–1918), geographer and philosopher/anarchist Piotr Kropotkin (1843–1921), and mathematician V. A. Kostitzin (1886–1963). This Russian paradigm, as it has been called, holds that ecosystems are the fundamental unit of life (as organisms cannot exist independently of ecosystems). Latvian evolutionary ecologist Edmund Lekevicius has explored this idea since at least 2002, following up on the early work of Ukrainian biologist Sergei Winogradsky (1856–1963) and the Russians mentioned. Watersheds have also been described as meta-organisms that evolve to increase the efficiency of infiltration and retention of moisture, mobilization and retention of nutrients, and drainage, mainly in the root zone. Hydrological systems have the capacity to adjust via selection processes, a topic we'll return to in Chapter 7.

The question of ecosystem evolution has often become ensnared in biologists' debates on the unit of selection. Does Darwinian selection, for instance, operate only on genes or also on phenotypes? Does it apply only to individuals or also to species? Some objections to the idea of ecosystem-level evolution can be reduced to a contention that ecosystems can't evolve because they don't have genes and because they contain multiple individuals of multiple species and therefore cannot be subject to evolution by natural selection. Okay, but such objections depend on a very narrow view of selection and evolution that, however relevant in some evolutionary biology problems, is irrelevant and obfuscating to landscape and ecosystem evolution.

Environmental or ecological filtering is also a form of selection, as environmental stresses and resource availability either prevent and counteract, or enable and support, establishment of certain biota. Selection occurs in purely abiotic, geophysical phenomena, too. The physical principle of least action, for example, allows and promotes some trajectories, structures and processes and excludes or inhibits others—a form of selection. *Homo sapiens* (and apparently some other species, such as ants that practice a form of cultivation) constantly practice selection by

preferentially utilizing higher-quality or more easily obtainable resources, as well as by purposeful modifications to meet needs
and desires. Gerald Nanson (Wollongong University) and He Qing
Huang (Chinese Academy of Sciences) in 2018 focused on fluvial systems, but their reasoning is applicable more generally:
evolution is possible in systems without genes or mutations. If
numerous variations exist with a wide range of functioning possibilities, these vary in terms of being advantageous for system
survival, and if there also exist selection mechanisms that favor advantageous variations, evolution is inevitable. And, as summarized
in Chapter 1 and discussed in detail in Chapter 6, more resistant, resilient, and efficient forms, structures, and patterns have
a greater probability of persistence and replication in landscapes,
encompassed by terms such as *efficiency* or *resistance selection*.
Natural selection narrowly conceived in the context of biological
evolution of organisms and species is highly relevant to ecosystem evolution—but so are all the other types of selection mentioned
above![9]

Paleoecologists have long accepted that ecosystems evolve,
using the term *ecosystem* to apply both to specific ecosystems
and to generic types (e.g., intertidal mudflats, tropical savannas).
Their work recognizes the interplay of abiotic change, species
evolution, and biotic-abiotic feedbacks. The paleoenvironmental
record, writ large, demonstrates that ecosystems change over time
and that both internal dynamics and external changes may drive
such change. It also shows that gradual environmental change
and disturbances can trigger ecosystem change and alter trajectories of ecosystem evolution. An example that is literally close to
(my) home is William Miller's work on the paleoecology of Quaternary transgressive stratigraphy in coastal North Carolina, including
the formations underlying my house (Figure 2.4). He showed that
as sea levels changed (driving changes in salinity, hydrology,
disturbance regimes, etc.) a phenomenon he called *community
replacement* occurred—that is, changes occurred at the level of
ecological communities as a whole rather than individuals. This

was true even with respect to specific habitats and niches, not just transition from, for example, nearshore to barrier island to estuarine environments. An additional example is a large body of work on grassland ecosystems linking evolution of plants, coevolution of plants, herbivores and soils, and triggers of change such as climate change, atmospheric chemistry, fire, and ecological interactions with woody plants.

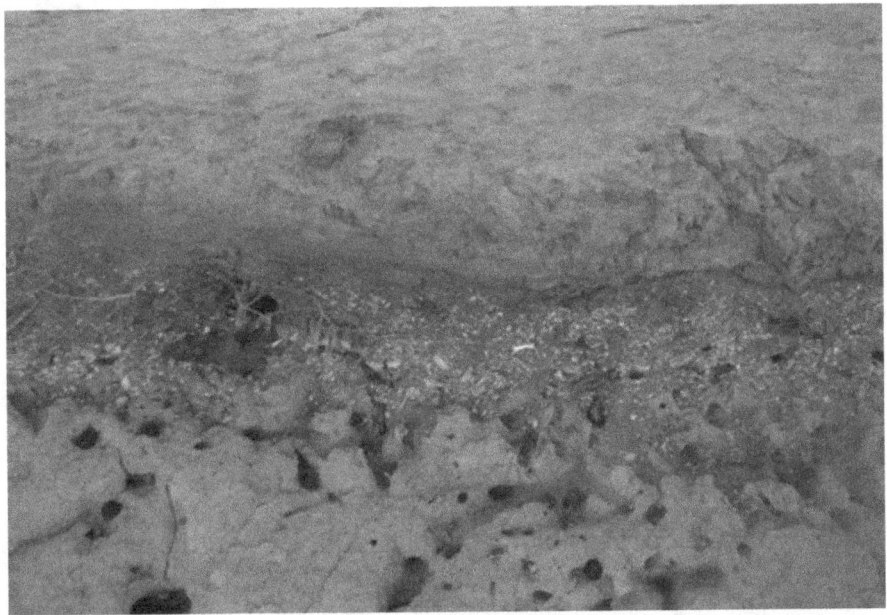

Figure 2.4. Shell layer of Flanner Beach formation, eastern North Carolina. William Miller used samples from shell layers like the one shown here to develop his community replacement theory. Area shown is about 1.5 m wide.

W. Miller, III, "Community Replacement in Estuarine Pleistocene Deposits of Eastern North Carolina," *Tulane Studies in Geology and Paleontology* 19 (1986): 97–122; W. Miller, III, "Paleoecology of Benthic Community Replacement," *Lethaia* 19 (1986): 225–231.

Another key thread is based on the idea that if rates of biogeochemical cycling or some other patterns or modes of resource use and flux confer some reproduction or survival advantages, then these are selected for, whether the advantages accrue at the level of individuals, communities, or ecological systems. Lapenis's biogeochemical selection ideas are one example, along with Belarussian climatologist Mikhail Budyko's (1920–2001) and hydrologist Peter Eagleson's (1928–2021) evidence that vegetation communities evolve to maximize water utilization. If more intensive resource use provides some advantage, then only two other simple assumptions (that resource availability is not being increased or decreased by external factors, and "biological saturation," which means that all niches are filled by something) are all that is necessary to dictate system-scale evolution along a path toward maximum rates and efficiency of resource use.[10]

Along yet another front, ecosystem-level evolution becomes clear in recent studies of plant-soil feedbacks, which connect ecosystem ecology with evolution. The phytotarium concept of Verboom and Pate mentioned earlier is one example, and arguments emerged earlier, in the late 1990s, that vegetation effects on soils can have evolutionary repercussions on plants (in addition to ecological filtering effects). Plant phenotypes respond to edaphic selection factors in soil that are in turn modified by traits of vegetation. Genetic variability of plant traits is linked to modifications of both pedological processes and soil microbial communities. These in turn influence vegetation phenotypes.

A gene-to-ecosystem conceptual model presented by Michael van Nuland (Stanford University) and colleagues in 2016 shows how both biological evolutionary and ecological processes in plant-soil feedbacks can influence—and even propel—both ecological change and evolutionary dynamics. The feedback relationships themselves can even evolve in different environmental settings. This builds ecosystem-scale loops: plants respond to soil microbes; plant-microbe interactions drive nutrient cycling, and these interactions vary along gradients in environmental state factors that are crucial for both natural selection and ecosystem processes. Figure 2.5 represents my take on the gene-to-ecosystem model, incorporating extended phenotypes. Work along these lines continues; van Nuland and company have shown how plants select soil microbial communities that direct the timing of their spring leaf emergence.[11]

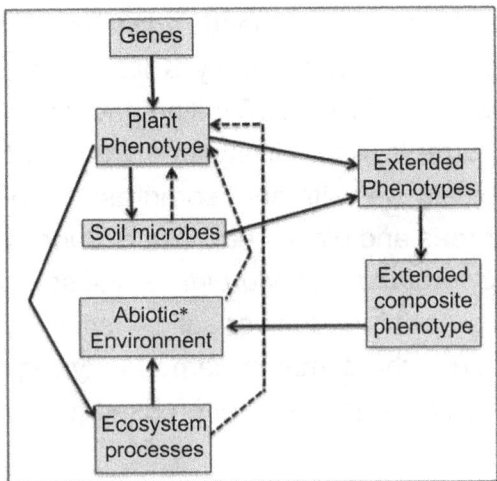

Figure 2.5. Gene-to-ecosystem model. Solid arrows indicate ecological change; dashed arrows indicate natural selection. Adapted with permission from J. D. Phillips, Landscape Evolution. (Amsterdam: Elsevier, 2021).

SUPRA-ORGANIC EVOLUTION

Two brief examples of supra-organic landscape evolution are reviewed below: North Carolina coastal marshes and forests with argillic (clay-rich) soil horizons.

SALT MARSH

Let's start with an intertidal mudflat, exposed to the air at low water and inundated at high water on a regular basis (where dominated by regular lunar tides, this would be twice a day in many cases). This is a mostly geophysical phenomenon involving the transport and deposition of sediment and the fluctuation of water levels. If it lasts long enough (that is, if physical forces don't erode it away), algae and microbes colonize it. The organisms themselves, and their exudates (slimy or sticky stuff), stabilize the sediment, increasing its resistance to erosion and transport. They also increase the carbon and nutrient stock of the muddy substrate, which, along with a greater tendency to stay put, allows plants to colonize. Which plants? That depends on the available propagules. In a given region, there are usually a limited number of candidates that can withstand regular flooding and un-flooding, bashing by waves and currents, and salinity. In my part of North Carolina, those would typically be saltmarsh cordgrass (*Sporobulus alterniflora*) in saltier areas and black needlerush (*Juncus roemerianus*), big cordgrass (*Sporobulus cynosuroides*), and saw-grass (*Cladium jamaicense*) in brackish and lower-salinity sites (I give the scientific names because the common names vary regionally, and the same name—e.g., saw-grass—is often used for different species) (Figure 2.6).

Figure 2.6. Black needlerush (*Juncus roemerianus*) dominated marsh in the North River estuary, North Carolina.

The plant roots further stabilize the substrate and provide their own exudates. A rhizosphere evolves, with new fungi and bacteria adapted to the changing conditions and existing symbiotically with the plants. The plants drop leaves and die, adding organic matter to the substrate which hosts its own microbial community of decomposers. The aboveground portion of the plant absorbs energy and slows flows and wave celerities when the marsh is flooded, promoting deposition of mineral sediment. A true soil—not just a sedimentary substrate—develops.

Organic matter accumulates from successive generations of plants and plant debris from elsewhere washed in by tides and storms, and mineral sediment is deposited. The marsh accretes and gets higher, and the soil becomes thicker. As the marsh elevation increases, it becomes flooded less often. Deposition thus decreases, providing a negative-feedback brake on the process. The hydrologic regime, soil chemistry and other soil properties change accordingly, allowing other plants to move in, and the supra-organic evolution of the entire landform, hydrology, soil, and ecological complex continues to evolve as an inseparable complex.

Now let us note that the story above, while common, is far from the only pathway that occurs. Marshes may develop from swamp forests due to flooding and salinity effects of sea-level rise, for instance. Counteracting the accretion is densification and compaction of the marsh soil/sediment due to gravitational settling, organic decomposition, and dewatering, as the mass of accumulating material squeezes out some of the water. Also counteracting the accretion is sea-level rise, which may itself be exacerbated or offset by subsidence or tectonic uplift. Any sequence can be, and often is, interrupted or reset by disturbance (e.g., overwash with sand or erosional scouring) during a hurricane, nor'easter, or other event (or by human actions). Marshes can also develop in other settings, such as interdune depressions or by rising sea levels encroaching on non-wetland environments. But in any event, the whole landscape evolves with inextricable interactions among biota, hydrology, and geomorphology.

FORESTS AND ARGILLIC HORIZONS

Argillic horizons are subsoil layers that are enriched in silicate clays. Soils with argillic horizons, by implication, are vertical texture contrast (VTC) soils, where coarser surficial horizons overlie finer-textured subsurface horizons (Figure 2.7). Such soils, also called duplex soils, are globally common. How and why do argillic (often designated as Bt) horizons and VTC soils develop? Ample evidence suggests that it sometimes involves supra-organic evolution of forest ecosystems.

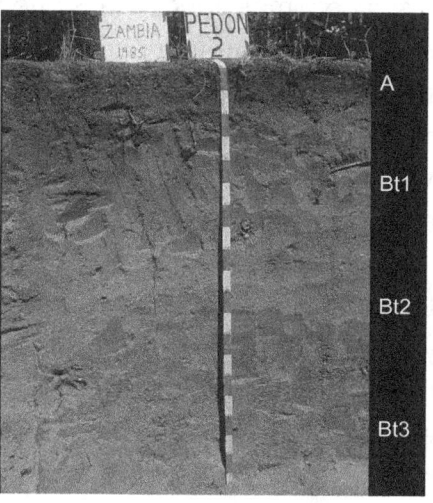

Figure 2.7. Multiple argillic horizons in a Kandiustult in Zambia. Photo by Stan Buol, reproduced with permission under a CC-BY 2.0 license.

The conventional explanation for VTC soils is that they form due to vertical translocation by percolating water. This water physically washes smaller particles out from between the larger ones (often sand grains) and moves them downward, concentrating smaller, clay-size material in the subsoil. This process is called *lessivage* or *argilluviation*. The water can also dissolve material from the upper layers that precipitate in the subsoil. Vertical translocation by water certainly occurs and is the single most important process for creating VTCs. But it is far from the only one! Preferential erosion of finer and/or deposition of coarser sediment at the surface can get the job done. Bioturbation often plays a key role, and texture contrasts can be partly or wholly inherited from parent material layering. In-place weathering and clay synthesis can produce silicate clays in subsoils, and upward movement of groundwater can lead to precipitation in the B horizon.

Three things spurred my interest. One was my own findings of variations in depth to the argillic horizon over short distances and small areas that could not be explained by any measurable variation in soil forming factors. This suggested to me that either something else in addition to vertical translocation by water was going on and/or that the translocation process is characterized by complex, perhaps even deterministically chaotic, dynamics. Second was a series of papers from 1987 on into the 2000s by Don Johnson and colleagues showing the strong but often, at least traditionally, overlooked role of organisms in creating and changing soil morphology.[12] The third was the book *Soils: A New Global View* by T. R. Paton, Geoff Humphries, and P. B. Mitchell (Yale University Press, 1995). In this book, they challenged the prevailing approach to pedology, favoring a more geological and geomorphological viewpoint. They also specifically challenged the notion of vertical translocation by water as a ubiquitous factor in soil formation. I did not, then or now, buy all their arguments, but they made a strong case for rethinking the conventional wisdom. So, from 1993 to nearly the present, I was involved in, and published several papers on, how VTC soils and weathering profiles are formed, complexity in pedogenesis, spatial variation of soils, and coevolution of soils, landforms, and ecosystems.

I came across an article by Verboom and Pate (2013) on ecosystem engineering of soils by eucalyptus trees in western Australia—including the synthesis of clays in the root zone! This led me to some of their earlier work showing that vertical redistribution of water and minerals dissolved therein resulted in formation of dense clay layers. The lateral root systems of the trees were, in essence, forming a subsoil clay layer by bringing the raw materials for clay synthesis into contact with each other. Because these clay-rich horizons benefit the trees via their water and nutrient storage, this is an example of positive ecosystem engineering. Thus, in at least one environment, trees and woodlands, if not forests *sensu stricto*, can build claypans and argillic horizons. Could this be a more general phenomenon?[13]

Around the same time, I recalled something I had read earlier in Greg Retallack's masterful *Soils of the Past* (3rd ed., John Wiley, 2019)—that soils with argillic horizons (Alfisols and Ultisols in the US Soil Taxonomy) do not appear in the paleosol (fossil soil) record until forests appeared in the Devonian. Coincidence? Unlikely. Deeper rooting depths of trees and effectiveness of weathering under forests likely play a role, but Retallack also noted the strongly tapering geometry of tree roots. These create large pathways for water movement in upper parts of the soil that taper down to nearly nothing at their tips. This allows translocated material to move downward but also to start building up at the end of the root line, so to speak (Figure 2.8). This is broadly consistent with my own work, which showed that tree roots (and root paths following death and decay) and insect and other faunal burrows are important in maintaining translocation, which might otherwise be reduced to negligible levels as low-permeability clays accumulate and as soil pores are blocked.[14]

Figure 2.8. Soil formed on limestone in Bohemia, Czech Republic, showing the close association between tree root depth and the argillic horizon.

Pierre Velde and Pierre Barre's (École Normale Supérieure, France) book *Soils, Plants, and Clay Minerals. Mineral and Biologic Interactions* (Springer, 2010) introduced to me a school of thought that I was not much aware of before: that soil clay minerals are fundamentally different from those formed below the soil or prior to soil formation by primarily geochemical processes. Independently of this perspective, however, they showed the role of plants and vegetation-based soil organic matter in the formation and retention of phyllosilicate clays, both directly and via their bacterial and fungal symbionts. Particularly important is plant uplift (via water intake) of silica and potassium, key building blocks of silicate clays. An important point is that clay minerals are a necessary source of the critical nutrient K (potassium), so facilitating the formation of clays that can retain K is of great benefit to the vegetation.

Without plants, Velde and Barre assert, there would be no clay accumulation in surface layers. However, their book does not directly address the "can trees make argillic horizons" question. First, they are concerned with plants in general, and grasses may be more effective clay formers than trees, at least when it comes to clay in A horizons. Second, much of their work is indeed concerned with A-horizon clay—it is in these layers that most root mass occurs, after all. They are less concerned with clay migration to the subsoil, though they do note that loss of clays from the surface layer is highly probable in forest soils. Finally, the book has more to say about retention of clays and clay minerals than the (flora-assisted) formation thereof.

In subtropical savannas of south Texas, there's a body of work on vegetation relationships to soils that doesn't quite fit into the trees-make-argillic horizons framework. Large woody patches occur on soils without argillic horizons, whereas smaller patches with more herbaceous vegetation are found in adjacent sites in the same landscape where an argillic horizon is present. This could occur because the clayey layers restrict root penetration, favoring more shallow-rooted grasses and shrubs rather than trees. One study showed that shrubs on argillic soils had less aboveground and greater belowground root mass than those on non-argillic soils. Root biomass and density on argillic soils was elevated at shallow (< 0.4 m) depths, whereas root density of the same species on non-argillic soils was skewed to depths > 0.4 m. Obvious relationships exist between the presence or absence of argillic horizons, root depth and biomass, vegetation-driven water use (including hydraulic lift), and soil hydrological properties, and similar results have been obtained in Australia. However, this body of work does not make it clear (at least to me) whether soil morphology is driving vegetation distributions, or vice versa, or both (via reciprocal interactions). It is also possible, I think, that the argillic horizons may be inherited from earlier, moister climates (this is sometimes the case even in desert aridisols) and poorly related to contemporary pedo-ecological dynamics with respect to argillic formation.[15]

Notwithstanding the Texas savannas, there is strong evidence that forests sometimes evolve as a supra-organic unit involving trees and other vegetation, soils, and their associated invertebrates and microbes. This has been documented in some cases,[16] and plant uplift of water, nutrients, and Si can help retain or even form silicate clays. The uplift is not restricted to trees or woody vegetation and does not necessarily result in subsoil clay concentrations. But as opposed to other plants, trees also provide or facilitate the processes to transfer clay to the subsoil, as illustrated by the fact that dominantly mineral forest soils (such as sandy podzols or spodosols), which lack the raw chemical materials for clay synthesis, show evidence of vertical translocation.

———————

To observe and study the seemingly mysterious ways of nature, we need not delve down to the realm of cells, particles, molecules, atoms, and the subatomic. Neither must we ascend to cosmic levels. They are manifested at the level of supra-organic ecosystems and landscapes. But even landscapes can exist, change, and be observed over a broad range of geographical extents and time spans. Thus, we must discuss the critical issue of *scale*, which we take on in the next chapter.

NOTES

1. Key references for the supra-organic ideas mentioned in this section: J. Brunetti, *The Farm as Ecosystem. Tapping Nature's Reservoir—Biology, Geology, Diversity* (Greeley, CO: Acres USA, 2014); F. E. Clements, *Plant succession. An analysis of the development of vegetation* (Washington, DC: Carnegie Institution, 1916); J. Lovelock, *The Ages of Gaia: A Biography of Our Living Earth* (New York: Norton, 1995); R. M. May, "Conceptual Aspects of the Quantification of the Extent of Biological Diversity," in *Biodiversity: Measurement and Estimation*, edited by D. L. Hawksworth, 13–20 (London: Chapman & Hall & The Royal Society, 1995); S. Rowe, What on Earth Is Life? An Ecological View. *Ecosystem Health* 7, no 3 (2001): 141–146; W. Jackson, *Consulting the Genius of the Place: An Ecological Approach to a New Agriculture* (Berkeley, CA: Counterpoint Press, 2010).

2. The interconnectedness of biota, geophysical and geochemical phenomena, landforms, and soils has long been recognized. Milestones include Dokuchaev's (1883) pioneering ideas of soils as the product of the combined influences of geology, climate, topography, and organisms, Vernadsky's (1926) biosphere concept and foundational work in biogeochemistry, and the ecosystem idea (Tansley, 1935). Dokuchaev (1883) did not explicitly deal with reciprocal interactions but did recognize that the factors of soil formation influence, and are influenced by, the other factors as well as the soil itself. The idea of ecosystems (and possibly the term) predates Tansley, but his 1935 article introduced the concept to a broad audience. For the most part, however, the subdisciplines mentioned here began to flower in terms of dedicated meetings, books, journals, and organizations from the late 1980s onward. References: V. V. Dokuchaev, "Russian Chernozem," in *Selected Works of V. V. Dokuchaev*, edited by N. Kaner (Jerusalem: International Program for Scientific Translations, 1883); A. G. Tansley, "The Use and Abuse of Vegetational Terms and Concepts," *Ecology* 16 (1935): 284–307; V. I. Vernadsky, The Biosphere, (New York: Copernicus, 1998). The original English edition was published in 1926.

3. Sources on soils as biomantles, phytotaria, and extended composite phenotypes: D. L. Johnson, "Biomantle Evolution and the Redistribution of Earth Materials and Artifacts," *Soil Science* 149 (1990): 84–102; J. S. Pate, and W. H. Verboom, "Contemporary Biogenic Formation of Clay Pavements by Eucalypts: Further Support for the Phytotarium Concept," *Annals of Botany* 103 (2009): 673–685; J. D. Phillips, "Soils as Extended Composite Phenotypes," *Geoderma* 149 (2009): 143–151; N. van Breemen, "Soils as Biotic Constructs Favouring Net Primary Productivity," *Geoderma* 57 (1993): 183–211; N. van Breemen and A. C. Finzi, "Plant-Soil Interactions: Ecological Aspects and Evolutionary Implications," *Biogeochemistry* 42 (1998): 1–19; W. H. Verboom and J. S. Pate, "Bioengineering of Soil Profiles in Semiarid Ecosystems: The "Phytotarium" Concept. A Review," *Plant & Soil* 289 (2006): 71–102; W. H. Verboom and J. S. Pate, "Exploring the

Biological Dimension to Pedogenesis with Emphasis on the Ecosystems, Soils and Landscapes of Southwestern Australia," *Geoderma* 211/212 (2013): 154–183.

4.   On landforms as extended phenotypes: J. D. Phillips, "Landforms as Extended Composite Phenotypes," *Earth Surface Processes and Landforms* 41 (2016): 16–26; D. Kim and K. Lee, "Landforms as combined expressions of Multiple, Reciprocally Interacting Species. Refining the Ecosystem Engineering Concept," *Earth-Science Reviews* 232 (2022): 104152. Examples of Dov Corenblit's and Woody Cotterill's contributions: D. Corenblit, et al., "Feedbacks between Geomorphology and Biota Controlling Earth Surface Processes and Landforms: A Review of Foundation Concepts and Current Understandings," *Earth-Science Reviews* 106, no. 3–4, (2011): 307–331: F. P. D. Cotterill and M. J. De Wit, "Geoecodynamics and the Kalahari Epeirogeny: Linking Its Genomic Record, Tree of Life, and Palimpsest into a Unified Narrative of Landscape Evolution," *South African Journal of Geology* 114 (2011): 489–514.

5.   See J. D. Phillips, "Biogeomorphology and Contingent Ecosystem Engineering in Karst Landscapes," *Progress in Physical Geography* 40 (2016): 503–526.

6.   R. J. Huggett, *Geoecology. An Evolutionary Approach* (London: Routledge, 1995); J. Major, "A Functional, Factorial Approach to Plant Ecology," *Ecology* 32 (1951): 392–412; F. Perring, "A Theoretical Approach to a Study of Chalk Grassland," *Journal of Ecology* 46 (1958): 665–679; J. D. Phillips, "An Evaluation of the State Factor Model of Soil Ecosystems," *Ecological Modelling* 45 (1989): 165–177; G. A. Pope, R. I. Dorn, and J. C. Dixon, "A New Conceptual Model for Understanding Geographical Variations in Weathering," *Annals of the Association of American Geographers* 85 (1995): 38–64.

7.   J. D. Phillips, "State Factor Analysis of Ecosystem Response to Climate Change," *Ecological Complexity* 40 (2019): 100,789.

8.   Ibid.

9.   G. C. Nanson and H. Q. Huang, "A Philosophy of Rivers: Equilibrium States, Channel Evolution, Teleomatic Change and Least Action Principle," *Geomorphology* 302 (2018): 3–19.

10. M. I. Budyko, *The Evolution of the Biosphere* (Boston, MA: D. Reidel, 1986); P. S. Eagleson, *Ecohydrology: Darwinian Expression of Vegetation Form & Function.* (New York: Cambridge University Press, 2002); J.D. Phillips, "Goal Functions in Ecosystem and Biosphere Evolution," *Progress in Physical Geography* 32 (2008): 51–64.

11. M. E. van Nuland et al., "Plant–Soil Feedbacks: Connecting Ecosystem Ecology and Evolution," *Functional Ecology* 30 (2016): 1032–1042; M. E. Van Nuland et al., "Natural Soil Microbiome Variation Affects Spring Foliar Phenology with Consequences for Plant Productivity and Climate-Driven Range Shifts," *New Phytologist* 232 (2021): 762–775.

12. For example. D. L. Johnson, E. A. Keller, and T. K. Rockwell, "Dynamic Pedogenesis: New Views on Some Key Soil Concepts and a Model for Interpreting Quaternary Soils," *Quaternary Research* 33 (1990): 306–319.

13. For example: Verboom and Pate, 2006; 2013 (endnote 4).

14. J. D. Phillips, "Formation of Texture Contrast Soils by a Combination of Bioturbation and Translocation," *Catena* 70 (2007): 92–104.

15. A. J. Midwood, T. W. Boutton, S. R. Archer, and S. E. Watts, "Water Use by Woody Plants on Contrasting Soils in a Savanna Parkland: Assessment with Delta H-2 and Delta O-18," *Plant and Soil* 205 (1998): 13–24; I. A. M. Yunusa et al., "Priming of Soil Structural and Hydrological Properties by Native Woody Species, Annual Crops, and a Permanent Pasture," *Australian Journal of Soil Research* 40 (2022): 207–219; Y. Zhou, T. W. Boutton, X. B. Xu, and C. H. Yang, "Spatial Heterogeneity of Subsurface Soil Texture Drives Landscape-Scale Patterns of Woody Patches in a Subtropical Savanna," Landscape Ecology 32 (2017): 915–929; Y. Zhou, S. E. Watts, T. W. Boutton, and S. R. Archer, "Root Density Distribution and Biomass Allocation of Co-Occurring Woody Plants on Contrasting Soils in a Subtropical Savanna Parkland," *Plant and Soil* 438 (2019): 263–269; C. B. Zou, P. W. Barnes, S. Archer, and C. R. McMurtry, "Soil Moisture Redistribution as a Mechanism of Facilitation in Savanna Tree-Shrub Clusters," *Oecologia* 145 (2005): 32–40.

16. See references in endnote 15; also M. A. Ibrahim and C. L. Burras, "Distribution and Origin of Argillic Horizons across Iowa—A Novel Hypothesis," *Soil Science Society of America Journal* 77 (2013): 580–590.

3

It Depends on the Scale

INTRODUCTION TO SCALE ISSUES

Think about some of the processes involved in the function and evolution of landscapes. Photosynthesis, respiration, and transpiration in plants, for instance, are driven by biochemical processes that operate at the molecular scale. They are manifested at levels ranging from cells to individual plants to plant communities to ecosystems, to entire biomes, and at the planetary scale in the form of global biogeochemical cycles. They can be measured and modeled at each of those scales, too—but even though the molecular scale biochemistry ultimately drives the processes, and the global-scale biogeochemistry encompasses it all, the tools necessary to observe, analyze, model, and understand them vary across that range of scales. Similar reasoning applies to temporal scale—some of the chemical reactions and matter and energy transformations can occur almost instantaneously, but the whole biogeochemical system coevolved (and continues to do so) over billions of years.

Mass transport rates in geomorphology can be measured in units of meters per second (m sec^{-1}; e.g., sediment transport by water or wind) all the way down to millimeters per year (mm yr^{-1}; tectonic or isostatic movements) and even millimeters per thousand years (mm ka^{-1} or m Ma^{-1}) for long-term average denudation rates. Mass fluxes can occur at the scale of individual grains of silt all the way up to continents. Not surprisingly, the tools, from conceptual

Mysterious Ways. Jonathan D. Phillips, Oxford University Press. © Oxford University Press (2025).
DOI: 10.1093/9780197755129.003.0003

models to governing equations to measurement and observation instruments, are different across that vast range of scales.

And so on.

The point is that landscapes are affected by phenomena operating at time spans ranging from instantaneous to planetary evolution (~5 Ga) and at spatial scales from molecular to global (Figure 3.1). An old joke in the landscape sciences is that the answer to every question is *it depends on the scale*. You can lump scale problems in three possibly overlapping categories: resolution, scale linkage, and scale contingency.

The *resolution* problem deals with the issue of the level of detail necessary to adequately observe and represent a given feature or process. This is, in turn, determined by the characteristic scales at which processes operate and features vary (and the size). Resolution is often characterized by components of *grain* and *extent*. Grain is the resolution of observation or sampling (e.g., sample interval, measurement frequency, pixel size, plot dimensions). Extent is the total area, volume, or time span considered. Resolution has actualist aspects with respect to the relationship between units of measurement, observation, sampling, or calculation, on one hand, and the real on-the-ground scale of processes and variation on the other hand. There also exist practical aspects in the form of trade-offs between the level of detail and the area or time span that can be feasibly represented. These are often reflected in time and costs of computation and processing. In my experience working with digital elevation models (DEMs), for example, analyzing the exact same area using a 1 m vs. a 10 m resolution DEM was associated with computer processing time of hours vs. seconds. And, though intuition tells us that more (and more detailed) data is better, it does not always work that way, even if we discount the practical issues such as processing time.

Scale linkage, the second major issue, alludes to the problems of connecting representations along that vast range of scales applicable to landscapes. How are we to link evolution of karst landscapes over the Quaternary (last 2.6 Ma) to the chemical kinetics of

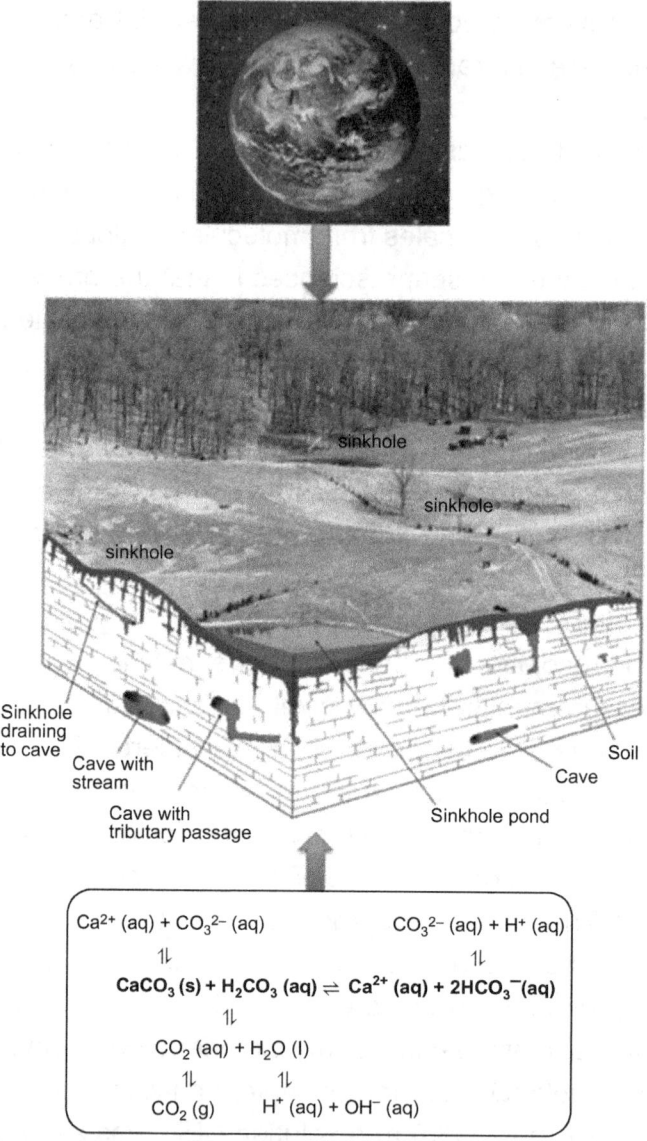

Figure 3.1. The karst landscape in Estill County, Kentucky (as depicted by the Kentucky Geological Survey in the middle image), like most geographical landscapes, is affected by a vast range of spatial and temporal scales. From the bottom, spatial scales as small as the molecular and as rapid as rates of chemical reactions and water flow are important. From the top, scales as large as global climate and as slow as geological evolution are important. Reproduced under the CC-BY 4.0 license from J. D. Phillips, "The Law of Scale Independence," Annals of GIS 28 (2022): 15–29, Figure 1.

dissolution of carbonate rocks (Figure 3.1)? How can we connect biome-scale transpiration to processes in individual leaf stomata? The scale linkage problem is that of functionally linking features and processes that operate and vary over different—sometimes vastly different—time and space scales. Operational scale linkage issues include identifying characteristic scales, deciding what should be included at a given scale, choosing or devising tools for transferring information and representations across scales, and determining conditions of scale independence (more on this below). Then there are theoretical issues of the extent to which rules, representations, or relationships are valid across the relevant range of scales, top-down vs. bottom-up influences, and multiple scale causality (MSC).

MSC refers to the fact that in the landscape sciences causality exists at multiple scales. The climate of a landscape, for example, is controlled by global atmospheric and ocean circulation, continental-scale patterns of air mass movements, regional factors such as topography and land/water juxtaposition, microclimatic patterns associated with land use and land cover, and patch-scale physics of the solar radiation energy balance. Geomorphology of mountain ranges is affected at scales ranging from global- and continental-scale plate tectonics to the forces acting on an individual grain of silt. This contrasts with laboratory and experimental sciences such as physics, where observed phenomena can often be taken as macroscale outcomes of microscale processes. For landscapes we typically observe system states created or influenced by multiple processes and controls operating at scales both smaller and larger than the scale of observation. In some branches of physics and chemistry, for instance, it is tacitly assumed that causality ultimately resides at the most detailed scale. A key implication of MSC is that the relevant "first principles" may reside at scales other than the smallest microscales.

The third of the overlapping general scale issues is *scale contingency*. The dominant controls over process-response relationships, as MSC tells us, often vary with scale. Scale linkage would be mainly a technical issue if the rules applicable to process-response relationships were constant across scales. These rules are typically not constant across scales, though, and empirical evidence, dynamical systems theory, and intuition all tell us so. Because the relevant representations and principles are contingent on spatial scale, then at some distance across the range of scales, the scales are independent of each other, in that what happens at one scale has no *direct* influence on what happens at the other. The forces acting on a grain of sand in a stream channel and geological scale development of the stream's drainage basin are both relevant to the evolution of the channel system. However, those grain-scale mechanics have no direct influence on the geological timescale and vice versa. In this and other cases, scale independence between scales near the ends of the spectrum of relevant scales is intuitively evident. However, in many cases, when the scales are closer together, independence is not so obvious. Scale contingency and linkage are intertwined and will be explored further below.

SCALE (IN)DEPENDENCE

Under what circumstances are scales independent of each other or not? The theory and mathematical basis of methods for assessing scale independence have been laid out elsewhere, but I will summarize their basic logic here from several different angles.[1]

Intuitively, we know that processes at the cellular and molecular level, for instance, though manifested through various intermediaries at the ecosystem scale, cannot directly explain ecosystem structures, functions, or evolution—or vice versa. But scale (in)dependence is not always obvious. In geosciences and ecology, scale linkage issues have long, and probably always, been at least implicitly recognized. Explicit consideration was spurred in 1965 by a couple of important articles. Peter Haggett of the University of Bristol articulated scale linkage problems in geography, and Stanley Schumm (1927–2011) and

Robert W. Lichty published a famous paper on the relationship between timescales and (in)dependence of variables and factors in geomorphology. Schumm and Lichty maintained that the same variables could be dependent or independent and that the same factors could be relevant or irrelevant at different timescales. Twenty-plus years later, Schumm and Lichty's arguments were given a formal mathematical basis and an explicit theoretical link to spatial as well as temporal scale.[2]

Space/time ratios are a direct way to think about and evaluate scale independence. These are based on comparing the spatial and temporal domains of the processes or changes involved. In the spatial domain, examples include ratios of distance or area. Uprooting of a single tree, for example, affects a ground surface area of a few square meters. Storm blowdown events, by contrast, often affect areas of hectares to thousands of hectares—therefore, areas of 10^0 to 10^1 m^2 vs. 10^4 to 10^7 m^2, suggesting scale independence. Conversely, in comparing uprooting of individual trees to the biomechanical effects of standing trees (for example by mass displacement of tree growth and infilling of stump depressions), the areas involved are not identical but are commensurate, and the scales are not independent with respect to those processes.

For temporal scale independence, we consider the rates, durations, and/or frequencies of the phenomenon and relaxation times. Relaxation time is the characteristic time it takes a landscape to respond to a disturbance or change. For example, groundwater and soil moisture flow velocities are typically measured in units of cm hr^{-1} or m day^{-1}. Surface flow velocities are measured in units of m sec^{-1}. This indicates scale independence, so that overland runoff is functionally independent of lateral seepage in the soil immediately underneath or that unconfined flow in groundwater conduits is independent of flow in the adjacent matrix (for example). Contrastingly, precipitation and evapotranspiration are generally measured using the same units (such as mm hr^{-1} or mm day^{-1}) and are not independent.

In 1979, Denys Brunsden and John Thornes (1940–2008, a geomorphologist in addition to being a TV meteorologist and author of a book of art criticism), two key thinkers with respect to issues to timescales in landscapes and then both at King's College, London, introduced the transient form ratio. This compares the relaxation time of a given event to the frequency of the event. A ratio >1 indicates a landscape dominated by transient forms that are typically unable to complete a response to a given disturbance. This was later generalized to a general tool for assessing potential scale independence, and the concept was reinvented by ecologists more than three decades later. One modification, where changes are continuous rather than discrete (sometimes referred to as press vs. pulse disturbances), used the duration of change rather than frequency of disturbance and a relaxation/duration ratio.[3]

Numerical climate models often rely on a computational tool called the Courant-Friedrichs-Lewy (CFL) criterion to ensure that changes are not propagated through model grid cells faster than they occur in time, which in turn ensures that numerical differing schemes are stable. This involves comparing the spatial and temporal resolutions—in climate models, grid cell size, and time steps—to make sure time steps are less than or equal to the grid spacing divided by the rate of the slowest process being modeled.

Outside the mathematical modeling realm, we are often stuck with spatial resolutions or time increments constrained by the measurement technology and data available. The CFL can be adapted to determine the appropriate time steps for faster and slower processes when the spatial resolution is fixed and the appropriate spatial resolution for cases where the temporal resolution is fixed. If results indicate unrealistically large, small, or different grid sizes or time steps, this signifies scale independence.

At a higher level of generalization *abstracted systems theory* proves the existence of scale independence, at least in a mathematical and systems theory context. First applied to ecological scale problems by the University of Arizona's William Schaffer in 1981, this construct applies to situations—most assuredly typical of landscapes and Earth surface systems (ESS)—where the whole nonlinear dynamical system includes interconnected elements operating at different scales. These can be divided into subsystems consisting of faster/slower (or larger/smaller) components and the interactions of the faster/smaller or slower/larger elements. Schaffer's product theorem for abstracted eigenvalues shows that when the scales of operation of the separated (abstracted) components are sufficiently different, they are independent of each other with respect to their effects on overall system behavior. "Sufficiently different" means at least two orders of magnitude or more (for instance, at least meters vs. hundreds of meters for length or years vs. centuries for time).[4]

Scale (in)dependence can also be approached via a technique called *entropy decomposition analysis*, based on entropy as a measure of uncertainty (and its converse, information). This shows that as the range of scale is increased by broadening or narrowing resolutions or by incorporating more controls, the influence of larger- or smaller-scale influences not only changes but may change qualitatively—e.g., in terms of having positive (entropy-increasing) or negative (information-increasing) effects. These qualitative causal shifts imply that a single causal explanation is unlikely to apply across the range of scales relevant to landscapes.

For those who may be interested, the mathematical details can be found elsewhere.[5]

For a simple example, consider shortleaf pine-bluestem (*Pinus echinata, Andropogon* spp.) woodlands in the Ouachita Mountains of southwestern Arkansas and eastern Oklahoma. This vegetation community is fire-dependent, with fires required at least once a decade to maintain them, and controlled burns are recommended every four to five years when used to maintain or restore shortleaf pine woodlands in the region. The timescale for forest development and tree maturity is about a century (recommended rotation ages for silviculture are 70 to 120 years). The difference in those timescales is not sufficient to consider forest landscape evolution and fire to be independent.

Trees—both individual plants and forest cover—have major impacts on landforms and soils in the Ouachitas (and elsewhere) due to their effects on runoff, erosion, sediment yields, bioturbation (i.e., soil mixing by organisms), weathering, and soil morphology. This occurs in a landscape characterized by complex, fascinating geology associated with an ancient tectonic collision zone and highly variable lithology. Between the trees and the rocks, we find scale independence—the very long timescales of geologic evolution are several orders of magnitude greater than those of the biogeomorphic effects of trees. Further, the geological processes affect areas of hundreds of square kilometers, while the area of influence of individual trees is a few square meters. In the vertical dimension, the tree effects are mostly confined to the upper 1.5 m (Figure 3.2), while the geologic processes may involve the entire crustal thickness.

Figure 3.2. Uprooted shortleaf pine, Ouachita National Forest, Arkansas.

To illustrate a case in the same region where scale independence is not so clear-cut, another occasional but severe disturbance in the area is tornadoes. The estimated recurrence interval of tornadoes in the Ouachitas is >2000 years, and these storms have profound impacts on forest ecology, soils, and landforms due to extensive uprooting and tree breakage (Figure 3.3). The recurrence interval of tornadoes (which is, by the way, getting shorter in this region of the United States, apparently due to climate change) is less than two orders of magnitude greater than the time frame of forest development. However, consider the spatial dimension: the estimated annual area of tornado disturbance in Polk County, Arkansas, in the heart of the Ouachitas, is 0.00017 percent of the land area. Scale independence seems a safe assumption.

Figure 3.3. Impacts of a mild (EF1) tornado, Ouachita National Forest, Arkansas.

In studying integrated evolution of forests, soils, and landforms in the Ouachitas, forest ecology, fire, and biogeomorphic (and pedologic) influences of trees and their interrelationships should all be incorporated. Tornado climatology and geological evolution, however, can be treated as boundary conditions or external influences.[6]

HIERARCHIES AND THE VANISHING POINT

What is the "vanishing point" at which relatedness between scales decreases to the point that the scales are independent? The abstracted systems argument indicates at least two orders of magnitude, and the various space/time ratios and entropy decomposition methods can be used to get at this. Yet another approach is through examination of scale hierarchies.

Scales of space and time are often arranged in hierarchies. Some are unambiguously additive, such as biological levels of individual, population, community, ecosystem, landscape, biome, and biosphere (Figure 3.4). Others are spatially nested functional hierarchies, such as that of hillslopes to zero-order drainage basins to first-order and n^{th} order watersheds. Nested spatial hierarchies can also be imposed by nested resolutions of maps or pixel sizes. Some landscape scale hierarchies based on conceptual models

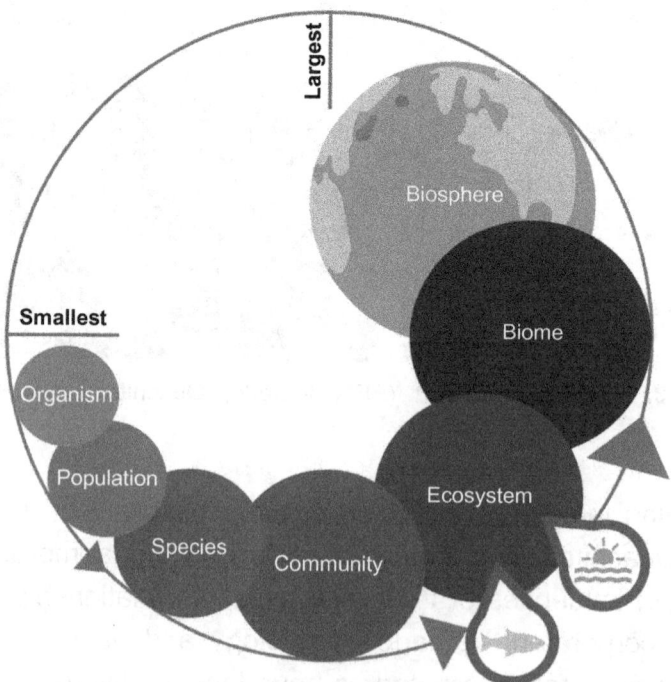

Figure 3.4. An example hierarchy—ecological levels of organization. Reproduced from Valeri Natole via pixabay.com.

may have less-distinct boundaries. But some of the latter are widely accepted and not controversial, such as the pedological hierarchy proposed by Dutch soil scientist J. C. Dijkerman in 1974.[7]

Hierarchy theory (HT) formalizes the notion that phenomena that operate or vary over similar scales are more closely related than those manifested at different scales. HT has been applied in geomorphology, ecology, pedology, and hydrology, though most often as a pedagogic or heuristic concept—but there are examples of analytical applications, too. HT assumes a nested structure of scales. Any particular level is directly affected by factors and dynamics operating at that level, at one level below, and at one level above. Scales two or more levels different from the scale of observation involve factors that are too slow or operate at too broad a scale, or too fast or at too fine a resolution to be observed. This does not mean there is a total lack of connection across the levels but that effects at the scale level of observation from scales more than one hierarchical level away are mediated by intermediate levels. Therefore, for instance, a fourth-order stream is directly influenced by the third-order streams flowing into it and the fifth-order stream it flows into. However, effects of second-order and smaller or sixth-order and higher streams are not directly observable. HT indicates that scale linkage must be stepwise along the hierarchy; it is not a tool enabling or a concept implying seamless linkage across the entire range of relevant scales; moving up or down the scale hierarchy, new factors and processes come into play and others become irrelevant.

Therefore, according to HT, scale independence exists between scales two levels or more apart. When the levels represent scales at least an order of magnitude different, the abstracted systems theory supports the assumptions of HT. Graph theory based analyses also support HT in this regard. Oregon State University physicist Juan Restrepo and coworkers in 2007 showed, using algebraic graph theory, that in systems involving slow components that encompass faster subcomponents (each represented as graphs or networks) the fast and slow components operate separately in terms of their effects on system coherence and complexity—a result directly analogous to the abstracted systems argument in this context. I did some analyses in 2016 applied directly to landscape hierarchies, with each level in the hierarchy treated as a subgraph of the overall system graph. This showed, for both generic models of ESS hierarchies and examples involving soils and fluviokarst landscapes (Figure 3.5), that graph synchronization—the extent to which different portions of the graph change contemporaneously or in sequential order—decays rapidly with hierarchical levels and becomes very low between nonadjacent levels.

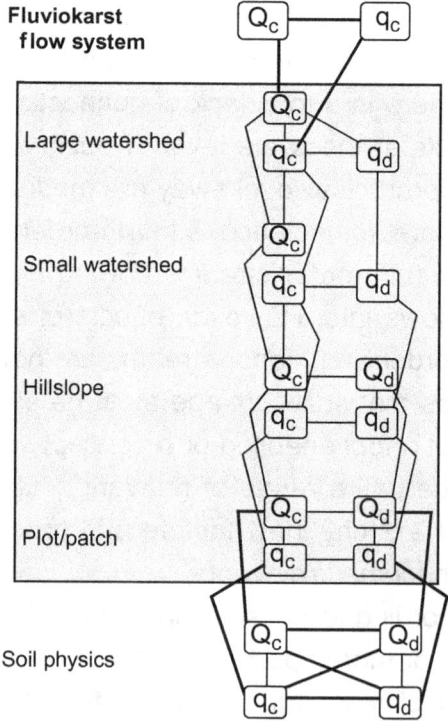

Figure 3.5. A hierarchy I used in studying the "vanishing point" in fluviokarst systems. Q and q refer to surface and subsurface flow, respectively, and the subscripts c and d to concentrated and diffuse flow. Reproduced with permission from J. D. Phillips, "Vanishing Point: Scale Independence in Geomorphic Hierarchies," Geomorphology 266 (2016): 66–74.

The graph analysis of the "vanishing point" in scale hierarchies indicates that relatedness declines with differences in scale or resolution, analogous to distance decay in the spatial domain. Scale linkage in landscapes and ESS is therefore best accomplished by identifying and focusing on the most important or interesting scale levels and moving stepwise from there rather than by attempting to identify the smallest- or largest-scale levels and working bottom-up or top-down.

A side note: *Vanishing Point* is also the title of an iconic 1971 movie directed by Richard Sarafian considered by many critics to capture the *zeitgeist* of the era (I was thirteen years old when it came out). When I first attempted to publish work on this topic, I attempted to pair it up with another of my favorites from that era, Hunter S. Thompson's book *Fear and Loathing in Las Vegas: A Savage Journey to the Heart of the American Dream*. However, the reviewers and editors of the journal *Geomorphology* weren't having it, and my original title (Vanishing Point: A Savage Journey to the Heart of Scale Contingency) had to be changed.[8]

REAL, REALIZED, AND ELAPSED TIME

Because of the long spans of time often involved in landscape evolution, we are frequently forced to resort to estimating rates of change or movement as virtual rates. Actual rates are derived from observation, measurement, or estimates of changes or processes during their actual occurrence. Virtual rates are based on the amount of change or movement over a time interval divided by the time increment. Thus, virtual rates are averages that often include periods of variable rates, not infrequently including lengthy periods of little or no activity. Thus, my actual rate of movement when I drive from New Bern, North Carolina, to Lexington, Kentucky, is measured by the speedometer on my 2010 Honda Civic and includes velocities that would get me a ticket if concurrently measured by the highway patrol, zero when I am at a rest stop or pumping gas, and everything in between—so, 0.0 to 128.8 km hr^{-1}. A virtual rate—say, 10 hours drive time divided by 956 km—is 95.6 km hr^{-1}. This is, of course, nothing more than a mean rate, and such averages can mask a lot of variation. An actual measurement of floodplain sedimentation, for instance, could be based on measurements of accumulation on marker horizons after every overbank flow event. A measurement of sediment thickness or mass divided by the time since deposition began (based, for example, on tree ring dating of partially buried trees or dating of basal deposits) gives a virtual, average rate. All the sediment could have been deposited in a single event, or (far less likely) at a consistent rate over time, or anything in between. There is no way of knowing if deposition occurred entirely or mainly at the beginning or the end of the period, and measurements could also represent a *net* accumulation (deposition minus erosion). In many cases, there is no help for this—there is no way to set up a measurement program for things that have already happened or to keep one going indefinitely. It is important, however, to bear the actual/virtual distinction in mind when assessing rates, durations, and timescales in landscapes.

Understanding landscape evolution involves at least three concepts of time: *real, realized,* and *elapsed* time. The rate or pace at which processes occur, such as the velocities of fluid flows or rates of biological metabolism or chemical reactions, are *real time.* Real time is associated with actual (vs. virtual) rates, although sometimes virtual rates can be a good approximation of the real. Actual, real-time rates of sublimation, evaporation, accumulation, and loss or refreezing of melt water on a glacier are difficult and expensive to measure, for example. Net gain or loss of mass over a year is a virtual rate but represents a reasonable approximation of real time with respect to glacial dynamics and glacial processes of landscape evolution (Figure 3.6).

The amount of time necessary for specific landscape changes to occur or for specific patterns and forms to emerge is *realized time.* Removal of forests by fire, logging, or pest infestations, for example, leads to increases in runoff, stream flow, and soil erosion.

Figure 3.6. Glacial ice in Greenland. Rates of melting, sublimation, snow input, etc. are difficult to measure. However, the virtual rate of annual mass gain or loss is an approximation appropriate for glacial dynamics. Photo by Richard Fletcher via Pexels.com, public domain.

In some landscapes, the changes are temporary, and within a few years reestablishment of vegetation cover returns the landscape to or near its pre-disturbance state (with respect to runoff and erosion). In the subtropics of the southeastern United States, for example, the realized time for hydrogeomorphic recovery is typically two or three years. However, the pedological realized time to rebuild forest soils and the ecological realized time to restore a mature forest will be much longer than for runoff and erosion. Realized time is therefore meaningful only with respect to specific phenomena. Realized times are strongly geographically and historically contingent. In the deforestation example, the environmental setting, the nature of the disturbance and response, characteristics of the adjacent and regional vegetation community, and timing (wet vs. dry years or winter vs. summer, for instance) all influence the realized time (many readers will recognize that realized time is closely related to relaxation time).

Elapsed time is the interval from the beginning of a period of change and the time of observation and is therefore related to virtual rates. It may indicate time since the very beginning of (an episode of) landscape evolution, such as the period since "time zero" of soil or landform development or from exposure of a substrate undergoing ecological succession. Methods of or constraints on measurement or observation may influence elapsed time. One example is tree ring based (dendrogeomorphic) estimates of erosion or deposition. Elapsed time is the interval from the establishment of a root or trunk and its age at the time of ring-counting. In chronosequence studies, elapsed time is the period represented by the chronosequence, which is in turn constrained by the ability to date different sites.

And speaking of realized time, I realize that it is time to move on to the take-home message of this chapter.

SCALE CONTINGENCY

It depends on the scale.

What depends on the scale? In landscape sciences, pretty much everything. Relevant processes operate at scales that range from planets down to molecules and occur over time spans ranging from billions of years to instantaneous. Scale contingency is therefore an intrinsic aspect of landscape evolution that cannot be avoided.

Resolution often determines what can or cannot be observed. This is observational or *epistemological scale contingency*. Rapid or small changes can be invisible if the time steps are too long or if the spatial resolution is too coarse. However, a broader, more expansive perspective may be required to detect some temporal and spatial patterns and features. Landscape features that are difficult or impossible to perceive on the ground are frequently detected in maps, aerial photographs, or satellite images. And of course, long-term trends and long-duration events can't be observed if the time window is shorter than the trend or duration.

There is also phenomenological or *ontological scale contingency* associated with the fact that the dominant processes and controls over landscapes vary with scale. Thus, what is known from, or about, one scale may or may not be applicable at another. Ontological scale dependency requires addressing issues of scale linkage and scale contingency.

The bottom line: landscape evolution is contingent—on time, place, and scale. This is important as we delve further into nature's mysterious ways.

NOTES

1. The full details are laid out in J. D. Phillips, "The Law of Scale Independence," *Annals of GIS* 28 (2022): 15–29

2. P. Haggett, "Scale Components in Geographical Problems," in *Frontiers in Geographical Teaching*, edited by P. Haggett and R. J. Chorley, 164–185 (London: Methuen, 1965); S. A. Schumm and R. W. Lichty, "Time, Space, and Causality in Geomorphology," *American Journal of Science* 263 (1965): 110–119; J. D. Phillips, "The Role of Spatial Scale in Geomorphic Systems," *Geographical Analysis* 20 (1988) 359–368.

3. D. Brunsden and J. B. Thornes, "Landscape Sensitivity and Change," *Transactions of the Institute of British Geographers* 4 (1979): 463–484; J. D. Phillips, "Humans as Geological Agents and the Question of Scale," *American Journal of Science* 297 (1997): 98–115.

4. W. M. Schaffer, "Ecological Abstraction: The Consequences of Reduced Dimensionality in Ecological Models," *Ecological Monographs* 51 (1981): 383–401.

5. See endnote 1.

6. Material on the Ouachitas is based on several years of research with Dan Marion of the USDA forest service (now retired), Professor Alice Turkington of the University of Kentucky, and other forest service scientists and several University of Kentucky students. References: J. D. Phillips et al., "Geomorphological Impacts of a Tornado Disturbance in a Subtropical Forest," *Catena* 125 (2015): 111–119; J. D. Phillips, D. A. Marion, and A. V. Turkington, "Pedologic and Geomorphic Impacts of a Tornado Blowdown Event in a Mixed Pine-Hardwood Forest," *Catena* 75 (2008): 278–287; J. D. Phillips, A. V. Turkington, and D. A. Marion, "Weathering and Vegetation Effects in Early Stages of Soil Formation," *Catena* 72 (2008): 21–28; J. D.

Phillips, and D. A. Marion, "The Biomechanical Effects of Trees on Soils and Regoliths: Beyond Treethrow," *Annals of the Association of American Geographers* 96 (2006): 233–247; J. D. Phillips, D. A. Marion, K. Luckow, and K. R. Adams, "Nonequilibrium Regolith Thickness in the Ouachita Mountains," *Journal of Geology* 113 (2005): 325–340; J. D. Phillips, K. Luckow, D. A. Marion, and K. R. Adams, "Rock Fragment Distributions and Regolith Evolution in the Ouachita Mountains," *Earth Surface Processes and Landforms* 30 (2005): 429–442; J. D. Phillips and D. A. Marion, "Biomechanical Effects, Lithological Variations, and Local Pedodiversity in Some Forest Soils of Arkansas," *Geoderma* 124 (2005): 73–89; J. D. Phillips and D. A. Marion, "Pedological Memory in Forest Soil Development," *Forest Ecology and Management* 188 (2004): 363–380.

7.   J. C. Dijkerman, "Pedology as a Science: The Role of Data, Models and Theories in the Study of Natural Soil Systems," *Geoderma* 11 (1974): 73–93.

8.   I was able to keep only part of the title: J. D. Phillips, "Vanishing Point: Scale Independence in Geomorphic Hierarchies," *Geomorphology* 266 (2016): 66–74.

4

Landscape Memory

LEGACIES OF A CRAPPY JOB

In 1975, in my junior (and as it turns out, final) year of high school, I was something of a hot mess. I was rebellious, obstinate, and had discovered—but not yet learned how to handle—alcohol and marijuana. I am embarrassed to say more, other than that I dropped out of high school. I took a factory job in a Cargill, Inc. egg-packing plant. It was shit work, and I hated it. My most common task was steam washing egg-shipping crates in an un-air-conditioned, barely ventilated, garage-like space in a subtropical summer. Otherwise, my job was to substitute for whomever might be on vacation or out sick.

The facility where I worked supplied eggs for all the Winn-Dixie grocery stores in North Carolina and Virginia, of which there were many in the 1970s. The most important and challenging job in the plant was at the end of a conveyor belt where crates packed with eggs (each crate a particular size and grade) were to be placed on a shipping pallet according to the orders from the grocery stores. Eight hours of lifting, carrying, checking pallets against a clipboard of orders, keeping track of what pallet went to which truck, and knowing that if you stepped away to take a leak or get a drink of water, dozens of crates would be backed up waiting for you on your return was exhausting.

Mysterious Ways. Jonathan D. Phillips, Oxford University Press. © Oxford University Press (2025).
DOI: 10.1093/9780197755129.003.0004

The guy who did that went on vacation. They replaced him with a seventeen-year-old kid who had only been at the plant a couple of months and who had never worked that station. The results were predictably disastrous, and the idiot foreman at the plant eventually had to assign three of us to do the job that one guy normally did. Another time, the same idiot plant foreman told me to drive a semi-truck and trailer (yes, an eighteen-wheeler) across town to a warehouse and back it up to a loading dock. Yes, really. At the time I could not even drive a stick shift, much less a tractor-trailer rig. I sat in the cab for a while, pondering the multiple gear shifts and other unfamiliar features. Finally common sense prevailed over the temptation to give them the truck wreck they deserved, and I declined the task.

Thus, my early experience as a high school dropout doing factory labor introduced me to corporate incompetence and the fact that one cannot necessarily rely on the judgment of bosses or superiors. I already knew about bad bosses and jerk bosses from jobs in the fast-food industry, but at least those guys knew how to run the grills, fryers, and cash registers, unlike the slack-jaw I was working for at Cargill.

I was making minimum wage, which I wasn't happy about, but I didn't expect any different from entry-level grunt work. What was eye-opening and disconcerting was that people who had been at the plant for five years, and whom I knew to be good workers, were only making twenty cents an hour more. Even the guy at the end of the conveyor belt, who by universal agreement was the best and most important person on the factory floor, wasn't doing much better.

Well, Cargill wrote the message on the wall for me in no uncertain terms. In that no-diploma working world, not only would I start out doing shit work for shit pay—which I expected—but I would also do so for much better-paid knuckleheads who had little idea what they were doing. Most importantly, it was clear that *it would never get any better.*

To hell with that. From dumping school, I transitioned rapidly to a desire to go to any school, anywhere, that would take me. I did well on the high school equivalency test, and Louisburg College, a church-affiliated junior college in my hometown, took a chance on me. I did well enough to transfer to Virginia Tech, and eventually the high school dropout went on to get a PhD.

So how does that relate to this book?

The answer is contingency. What if, after dropping out, I had gone to work at a job that was even half decent? What if I'd landed in a spot where I at least respected my bosses' professional competence? Where there was even a small chance of small advancements (and believe me, my ambitions were modest then)? There would have been a good chance, based on my memories of my seventeen-year-old self, that I would not have gone back to school. The whole trajectory of my life would have changed.

The same is true of landscapes. Chance events, whose effects are sometimes clear in retrospect, but which could not have been foreseen, can initiate or change the path of evolution. Even if the lousy job and the knucklehead boss had slipped my mind entirely, my life history since then and my current situation would "remember" them in the sense that the effects of the job are evident. That is an analogy to one form of landscape memory.

LANDSCAPE MEMORY

Landscape memory in this chapter is not about my or our memories of landscapes, but about what landscapes "remember" about the past. Landscape evolution takes time and is *historically contingent*—that is, the present (or any given slice of time) is influenced by the past. If things were otherwise, if evolutionary pathways were deterministic, or entirely predictable based on general laws and principles, landscapes could be understood without considering historical contingency. Landscape or other scientific studies based strictly on laws are ahistorical, as generalities about, say, thermodynamics, mechanics, and chemical reactions are independent of time and place. But landscapes are historically contingent and path dependent. This is manifested in several different ways.

The most clear-cut form of historical contingency is *inheritance*. Some properties and features persist for long periods, well beyond the processes and conditions under which they formed and are therefore inherited. In geomorphology, inheritance takes the form of resistant, durable features that are worn away but slowly. These are landform traits passed on as landscapes evolve. In fact, Rowl Twidale (University of Adelaide), who has perhaps done more research than anyone on inherited and ancient features in modern landscapes, uses the term *lineage* for cases where inherited features not only persist, but have discernible effects on subsequent landscape development. One of Twidale's examples is etch forms, formed by chemical weathering at the weathering front at the base of the regolith and then subsequently exposed when the overlying soil and regolith is eroded away (Figure 4.1). The cause of the unequal weathering at the weathering front, later showing up as inherited topography at the surface, is often rooted in ancient geological events. Twidale showed that causative histories of some such landforms may data back even to the Precambrian, the oldest (and longest) period of Earth's history. Other common examples, particularly at high latitudes and altitudes, are features associated with glacial and periglacial processes that no longer operate (Figure 4.2).

Inheritance is the key to the use of soils and paleosols as paleoenvironmental indicators in archaeology, paleoecology, geomorphology, historical geology, and paleoclimatology. This is based on an inversion of state factor logic—as soils are products of the environment, their inherited features contain information about the environment in which they formed (Figure 4.3). Victor Targulian and Sergey Goryachkin's (of Russia's Institute of Geography) concept of *soil memory*, first published in English in 2004, is based on inheritance. Soil memory was defined as the capacity of soils

Figure 4.1. Etch surface in Australia's Flinders Ranges. The etch surface is exposed on the slope in the foreground and remains buried further downslope.

Figure 4.2. Inherited landforms—rocky talus slope, stone circles, and (in the background) a glacially carved valley, Karkonosze National Park, Poland. These features are inherited from periglacial and glacial process regimes in the Pleistocene.

for storing information about environmental factors and pedogenic processes (other concepts of soil or pedological memory exist; more on that below). In this instance, soil memory is the unique set of soil properties that are inherited and that reflect the interactions of the soil-forming factors. Targulian and M. A. Bronnikova in 2019 linked soil memory to the broader concept of geosystem memory.[1]

Figure 4.3. Paleosol at the base of the Flanner Beach formation in eastern North Carolina. The gray soil was formed from mud deposited in a wet environment, and the dark woody material is from the roots of bald cypress (Taxodium distichum). This reveals that about 700 Ka this was a freshwater swamp.

Inheritance is also important in ecology, though the typical time spans are shorter. *Biological legacies* are defined as living organisms, seeds, and organic debris in soils and substrates and also include biologically derived patterns (such as root channels or faunal burrows). Inherited biological legacies influence future ecological succession and soil formation. The type and extent of biological legacies inherited after disturbances may produce persisting ecological landscape patterns, a phenomenon that has been called *ecological memory* (Figure 4.4). For example, in pyro-phytic (vegetation adapted to and shaped by fire) shrublands, root stock from pyrophytic woody plants that survives a blaze maintains pyrophytic shrublands. By contrast, if the post-burn inheritance is pyrophobic (fire intolerant and nonresistant) seed stock, pyrophobic forests develop. Inheritance in the form of organisms (or propagules) that survive disturbances strongly affects ecosystem development in many other cases, too, and is related to resistance selection (discussed in Chapter 6). Ecological memory is also relevant at millennial timescales, as paleoecologists have shown.[2]

Figure 4.4. Ecological memory, Podyji National Park, Czech Republic. The shovel is in a depression created by rotting of the stump of an oak tree. The three nearby trees grew from root stock of the original tree, upon which coppicing was practiced.

In natural selection, inheritance is obvious in the heritability of traits but is also present at the scale of genotypes. Lekevicius notes that diversity of genotypes (which is maintained by abiotic as well as biotic factors) "is a kind of population memory. Selection must have saved information about different environmental conditions . . . encountered in the past." Therefore, the gene pool of these populations is prepared to respond to environmental changes that are within the range of those that have historically occurred.[3]

Another form of historical contingency is *conditionality*, which happens when development pathways vary according to whether specific circumstances exist. A railroad switch is a simple metaphor for conditionality, where the train's path is guided by the position of the switch (Figure 4.5).

Figure 4.5. Railroad switch: a simple but accurate metaphor for conditionality. Courtesy of pxhere.com.

Conditionality in Earth surface systems (ESS) is most readily seen in ecological succession and ecosystem evolution. Disturbances are a common example. Whether or not a landscape has been burned, flooded, logged, or farmed, for instance, can set the path of evolution. Studies of vegetation change—whether in natural or human-altered or disturbed or undisturbed settings—often show that development of vegetation communities is unpredictable. This includes studies of sites with similar environments, so that place factors are as consistent as possible. As laws are by definition consistent, this leaves history factors in the form of conditionality to explain the variability. Conditionality is evident in "founder effects" (which species happen to be the earliest arrivals) in determining the course of succession and is a major component of ecological

assembly theory, going back at least to the American ecologist
H. A. Gleason (1882–1975) in 1939.

Forks in the evolutionary pathways of pedological and geomorphological phenomena also abound. In Figure 4.6, for instance,
subsequent development of the recent sandy deposit is contingent on when the next storm arrives and whether it rearranges,
removes, or adds to the sand. Other key conditionalities include
survival of any buried plants, seeds, propagules that may be included in the deposit and the deposition of any fine sediments
along with the sand. I have observed cases of similar sandy deposits in the same vicinity either becoming stabilized by vegetation
and beginning to develop podzolic soil features within a few years,
disappearing within a matter of weeks, or encroaching further on
adjacent marshes.

Figure 4.6. Recent storm deposits along the Neuse River estuary shoreline,
North Carolina. Depending on the timing of subsequent storms, whether they
remove or add to the sand lobe, edaphic conditions for vegetation establishment,
the availability and viability of propagules and rootstock, post-storm development
could take several different paths.

EVOLUTIONARY PATHWAYS

Stuart Lane (University of Lausanne) and Keith Richards (Cambridge) in 1997 noted that development trajectories of rivers

and other geomorphic systems are path dependent, and that changes can be understood only by considering different sequences of changes in system states rather than via deterministic pathways. Subsequent work, particularly in fluvial systems, has supported their view. In 2015, Keith Beven (University of Lancaster) described conditionality in hydrological, geomorphological, and ecological systems in terms of the current state as a "nonlinear superposition of past states." On arid alluvial fans, Kathleen Parker (University of Georgia) in the mid-1990s documented conditionality whereby geomorphic processes lead to different soil types, which in turn lead to differences in vegetation. In the other causal direction, consider the case of biomechanical effects of trees, particularly with respect to whether they break off or uproot at their demise. In this case, vegetation changes lead to soil and geomorphic change. The studies I am most familiar with are those I have been involved in, with Dan Marion (US Forest Service) in Arkansas and with Pavel Šamonil (Sylva Tarouca Institute, Brno) and others in central Europe. These conditionalities have fundamentally different geomorphic and pedological impacts, at the scale of forest stands and of individual trees.[4]

The conditionality form of historical contingency is also intrinsic to many state-and-transition models (STMs) in ecology, pedology, and geomorphology, as state transitions depend on the occurrence of specific disturbances, management strategies, or environmental changes.

Bailey effects are named for George Bailey, the main character in Frank Capra's 1946 film *It's a Wonderful Life* (based on a story by Philip Van Doren Stern). Mr. Bailey is wrought with despair and attempts suicide after a run of bad luck, believing that his life has not been productive or worthwhile and that the world would be better off without him. A guardian angel intervenes, and Bailey is given the opportunity to view alternative evolutionary trajectories, as it were, to see what his community would be like if he had never been born. The message is that seemingly minor, insignificant actions of an individual may have ripple effects and chain reactions that produce dramatically different outcomes. In 2004, I proposed Bailey effects as a metaphor for interconnections based on conditionality. Stephen J. Gould of Harvard (1941–2002) earlier used the same metaphor in discussing contingency in biological evolution; his 1990 book *Wonderful Life* is a direct nod to the film.

Bailey effects are related to a third general type of historical contingency and landscape memory: *dynamical instability.* If (or when) a system is dynamically unstable, it is sensitive to small disturbances and/or minor variations in initial conditions. Therefore, effects of those disturbances and initial variations are magnified or amplified and become large and long-lived compared to the size and duration of the source. Dynamical instability means that conditionalities and inherited traits do not have to be large, or linked to major factors, to create alternative historical trajectories. Dynamical instability (in most cases identical to deterministic chaos) is quite common in landscapes, though certainly not all landscape systems are dynamically unstable, and those that can be unstable may not be at all times or at all scales.

Inheritance, conditionality, dynamical instability, and Bailey effects are overlapping categories, and the point is to highlight the major forms of historical contingency rather than to exclusively classify all situations. Memory in landscape evolution sometimes reflects the effects of factors no longer evident (but remembered in the landscape) and sometimes includes keepsakes such as inherited forms, artefacts, and fossils.

Looking back, to infer or reconstruct the history of a landscape, we are faced with the fact that from a starting point that may be known or unknown, many evolutionary pathways were possible— but only one historical trajectory actually occurred. Looking ahead, again we face multiple possibilities. Our task in prediction is to rule out the impossible, to constrain the possible by determining what is less or more probable, and to identify potential conditionalities.

SUCCESSION AND STATE TRANSITIONS

Succession is the most straightforward and commonly used model of ecological system development, usually these days without the baggage of communities as superorganisms and disturbance as non-normative events associated with Clementsian succession. Communities and ecosystems indeed do often develop systematically via a sequence of predictable steps toward a stable climax in real landscapes. Succession is also consistent with traditions that resolve any historical pattern into a sequence of stages, which is a perfectly logical approach in both scholarship and storytelling. Development by stages also seems consistent with human psychology and experience. The potential problems arise when we view successional pathways (or evolutionary trajectories more generally) as prescribed routes to be repeatedly followed rather than idiosyncratic paths that may or may not be repeated.

There is more than one way to interpret a sequence of changes in plant community composition, however. Where different locations exhibit similar sequences, this can be due to conceptual model equifinality. Equifinality is when different processes or controlling factors lead to similar outcomes. In ecological succession, the same sequences and outcomes can in many cases be described (and thus apparently explained) based on several different concepts of succession, as well as Gleason's plant sociology approach or assembly theory. Another example of model equifinality is topographic evolution, where the same landscape can sometimes be explained through cyclic/successional, progress-toward-equilibrium or -nonequilibrium lenses, as Bill Renwick of Miami University showed.[5]

Conceptual model equifinality is not necessarily bad, however. First off, it is inevitable, as we seek to concoct plausible explanations for our observations. Rare is the case where, at least early on, there are not multiple potential explanations to account for what exists. Often several of these pass tests of logic, consistence with known laws, and comparison with observations and data. This elevates them to competing hypotheses that can be subjected to more rigorous tests, or to criteria such as Ockham's razor—when two or more explanations fit the evidence equally well, the simplest is preferred. And, of course, even when a phenomenon has been well studied, there is often no single correct explanation or model, even for an individual landscape. In the landscape sciences, it has often been the case that two apparently competing explanations emerge, along with debates about which is correct. As more evidence accrues, the answer turns often out to be either or both (and occasionally neither).

For instance, there once existed competing schools of thought arguing that karst caves in carbonate rock form in the vadose/unsaturated zone, in the phreatic/saturated zone, or along the boundary between saturated and unsaturated conditions. We now know that formation above, below, or at the top of the saturated zone occurs, often contemporaneously within the same cave. Similarly, alternative theories of barrier island formation involved buildup of nearshore bars, drowning of coastal dune systems by the sea, and cutoff of spits across the mouths of estuaries. Again, we now understand that all three mechanisms occur and that some barrier islands reflect more than one of these. For a final example, there is the phenomenon of desertification, whereby semi-arid and subhumid forests, woodlands, savannas, and grasslands are turned into deserts. Desertification studies were once framed in terms of natural (or at least not directly anthropic) causes such as climate change and droughts versus human causes such as overgrazing, deforestation, and groundwater withdrawal. Within those two broad categories there existed sub-controversies over the relative importance of specific climate or human land-use phenomena. We came to understand that all the factors mentioned above, and more, may contribute to semi-arid land degradation (which became the preferred term for the phenomenon). The debates and uncertainties persist, but they now relate to specific regions or situations rather than to a purported single dominant driver.

The bottom line: multiple causes or pathways may lead to (or away from) similar landscapes.

As noted earlier, dissatisfaction with single-path sequential frameworks led to conceptual models that accommodate variable and more complex patterns—STMs, some of which are not necessarily explicitly labeled as such. This represents a general conceptual progression from fixed sequences and linear chronologies to state transitions and networks of potential changes. STMs are also amenable to applications of network-based and graph theory approaches to landscape evolution, as I pursue in ~~mind-numbing~~ fascinating detail in *Landscape Evolution*.

PATH DEPENDENCE AND CANALIZATION

In early stages of landscape evolution, there may be many, many possibilities for where, say, shrub patches, sinkholes, stream channels, or a forest blowdown may occur. Once these things happen, however, subsequent pathways where these features or events do *not* occur, or where they form in different locations, are foreclosed. This is *canalization*, which refers to situations whereby once an evolutionary path has been established, other previously possible pathways are eliminated, and the evolutionary trajectory is channeled, or canalized. British biologist C. H. Waddington coined the term in 1942 with respect to evolutionary genetics, where the term is most often used (other than, of course, the literal definition of canal building and river straightening). Canalization of evolution and development has since been applied more broadly to geophysical and ecological systems.

For instance, where a channel begins forming may be affected by such tiny variations that it is apparently random (due to dynamical instability). Once the channel is started, though, it is a more efficient flow path than adjacent areas and outcompetes them. The flow in the channel helps maintain it and may enlarge it by scour. The initial location of the channel was not necessarily dictated by any control, nor was it deterministically predictable. Once developed, however, some previously possible channel locations were excluded, and some down-gradient channel locations are made more or less probable (or eliminated). This example, where the channelization/canalization metaphor is literally applicable, illustrates the role of instability and positive feedback in canalization. This is evident in many other cases, such as karst landform evolution. Karst development features critical self-reinforcing feedbacks among weathering, dissolutional enlargement of conduits, cavities, depressions, and water flows.

Canalization is also strongly influenced by irreversibility. Weathering, erosion, decomposition, combustion, and other processes are irreversible and therefore contribute to canalization. For example, a decomposed log cannot be reconstituted, and the pool of carbon and nutrients beneath it guides soil changes, microbial ecology, and eventual establishment of new vegetation. A mud flow or other mass wasting event moves mass downslope that cannot move back up again, and the resulting changes in slope morphology influence future slope processes, making some subsequent developmental paths less or more likely (Figure 4.7).

Figure 4.7. Rockslide in Kachin State, Myanmar. The mass movement is irreversible and affects future landscape evolution, eliminating possible pathways that could have occurred with no slope failure. Photo: India Blooms News Service via maxpixel.net.

TYPES OF EVOLUTIONARY PATHWAYS

Let's get meta-. That is, rather than considering possibilities and probabilities of specific processes, events, and features, here we think about the extent to which evolutionary trajectories—chains of processes, events, and features—are inhibited or foreclosed, vs. triggered or encouraged, as landscape evolution rolls on. I have emphasized that landscape evolution only occasionally follows inevitable, predictable paths toward a singular outcome. Multiple possible outcomes and pathways are not the exception, but the rule. Yet, over years of studying these multiple-pathway systems, I started to wonder why we do not observe even more pathways. Given the extensive geographical variation in place factors, the influence of local history factors, conditionality, dynamical instability, Bailey effects, and so on, why are some seemingly possible landscape evolution histories and trajectories rarely or never observed?

It is hardly news that evolutionary trajectories differ with respect to their frequency and probability, but a general synthesis of evolutionary pathways of ESS will become increasingly important and useful as new challenges, such as climate and land-use change, continue to arise. The key question is why some historical sequences are common while others are rare. Some are apparently nonrecurring, one-off affairs, while others are reinforced.

If we start with the population of all imaginable evolutionary trajectories, we can break these down into eight different groups with respect to their possibility, probability, and common or uncommon occurrence: impossible, currently impossible, improbable, inhibited, transient, non-repeatable recurring, repeatable recurring, and selected pathways.[6]

Some imaginable trajectories turn out to be *impossible paths* because they violate generally applicable laws, or necessary conditions do not occur. For instance, pathways that would require energy transformations that violate conservation laws or upslope movement of gravity-driven flows can be ruled out. Impossible also are pathways that might be possible in general but not in context. In a tropical setting, for example, freeze-thaw phenomena are impossible. *Currently impossible pathways* have, or could have, occurred in the past, but they depend on boundary conditions that no longer exist. There exist some paleosols, for instance, that have no modern analogs, as their properties reflect the effects of extinct biota with no contemporary equivalents or an atmospheric composition that no longer exists. Ecological succession, pedogenesis, and geomorphic evolution linked to, for example, a glacial climate, cannot recur in currently unglaciated zones (at least until a new glaciation occurs).

Improbable paths for landscape evolution are possible but are infrequent or unlikely because the necessary conditions involve infrequent events or rare combinations of boundary conditions. Pathways that reflect responses to very rare flood or storm events or the formation and modification of impact craters from extraterrestrial objects are examples. Also possible but rare are *inhibited paths*. These are due to feedback mechanisms or resistance factors that impede development along the pathway (as opposed to their being rare because of dependence on unusual, infrequent boundary conditions or events). Once the non-native kudzu vine (*Pueraria montana*) is established on hillsides in the southeastern United States, other vegetation succession paths are forestalled, for instance, as kudzu quickly comes to dominate the vegetation and is nearly impossible to outcompete and exceedingly difficult to remove. Another example is the formation of clay-rich illuvial argillic horizons in soils. While these horizons can deteriorate, they are persistent and highly resistant, making pathways involving destruction of argillic horizons uncommon. Similarly, weathering may form secondary silicate and sesquioxide minerals that can be destroyed or leached away, but the resistance of these minerals makes pathways involving their removal rare. In many bedrock-controlled streams, for another example, resistant banks prevent significant bank erosion and lateral channel migration, so that fluvial adjustments are primarily in the vertical dimension. The link between these examples of inhibited pathways and canalization should be clear.

Negative feedbacks can prevent or impede some potential pathways by operating to maintain or restore certain system states. When vegetation is removed from humid subtropical forests of the southeastern United States, for example, it might develop along a pathway toward severe erosion, surface exposure of clayey subsoils, and sometimes extensive gullying—and this has occurred historically when land-use practices (such as farming) inhibited natural revegetation. If spontaneous plant regrowth is not inhibited, however, vegetation cover is typically quickly re-established, as is the pre-forest clearance runoff and erosion regime. This vegetation response is a negative feedback that impedes the pathway toward severe erosion.

Inhibition of development along a pathway is sometimes controlled by thresholds. Examples include carbonate formation that inhibits clay development and clay mineral compositions that impede bioturbation. However, the impediments do not occur if the thresholds of carbonate leaching and clay mineral ratios are not exceeded.

Some landscape evolutionary pathways are short-lived (the definition depends on the timescale of interest) and poorly preserved. These are *transient paths*. They may be common, but their limited persistence and duration makes their observation uncommon. An over-steepened slope is a good example of a transient pathway. The steepest angle that can be maintained in a material is called the angle of repose. If mass movements, erosion, sedimentation, landscaping, or other processes cause the slope to become steeper than this angle, the slope becomes unstable. Mass movements occur, and the gradient recovers to the angle of repose. Therefore, pathways involving such over-steepening are short. Another example is when weathering in humid environments produces soluble salts, which are then rapidly leached away.

Intrinsically dynamic, rapidly changing environments such as sandy coastlines are often characterized by transience. Responses to storms and other high-energy events, as well as multiple daily tidal cycles, are rapid and result in formation and swift destruction of features such as beach ripples and cusps and longshore bars and troughs. Transient paths typically produce outcomes that are unstable, fragile, or subject to reversal by offsetting processes. Transience may depend on the time frame of interest, as do repeatability and recurrence, discussed below.

Non-Repeatable Recurring Paths

There also exist evolutionary pathways that recur in different locations but cannot be repeated in a particular landscape on account of self-limitations or irreversible processes. Rock breakdown by weathering is irreversible, and chemical weathering depends on availability of weatherable minerals. Pathways including chemical rock breakdown and weatherable mineral depletion are common— but though they appear in numerous locations, they cannot be repeated at a specific location. Landscape denudation involves irreversible processes and is thus non-repeatable. These examples show how some paths are unrepeatable due to material limitations. The weathering of a granitic rock mass or leaching of salts and carbonates from soil in a humid climate, for instance, is a one-way street that can be traveled only once.

In other cases, paths that occur in multiple, even numerous, different locations are non-repeating in any one landscape because they encounter some sort of ultimate limit. A downcutting stream can incise until it reaches its base level. But a downcutting pathway cannot recur in the same stream without some lowering of base level downstream or rejuvenation by uplift upstream, for example.

Repeatable Recurring Paths

By contrast, some recurring paths may be repeated in the same setting. These pathways are not inhibited or self-limited and may be renewed by recurring or ongoing conditions or processes. This category includes most types of ecological succession, particularly in landscapes with recurring disturbance such as fire, grazing, or storm overwash. Evolutionary pathways of growth, extension, and cutoff of river meanders in laterally migrating alluvial channels are also potentially repeatable. Other recurring paths are cyclical. One example in some rivers occurs as floodplains accrete and build higher and higher banks, confining larger and larger flows. Eventually a major flood or series of floods strips away the accreted alluvium, which begins the sequence anew. Removal of surface soil by erosion may stimulate production of new soil by bringing the bedrock weathering front closer to water inputs and biological activity at the surface, thus increasing weathering. The sequence can be repeated later after another erosional episode or period. Some landscape developmental pathways and processes and pathways are linked to ongoing or seasonal biological inputs such as flowering, leaf-out, litterfall, organic matter decomposition, and humification and are thus recurring (though at some timescales these may be considered transient).

Selected Paths

Recurring pathways, repeatable or otherwise, are often *selected paths* that are preferentially developed, enhanced, or preserved by feedbacks (discussed further in Chapter 6). In geomorphic, pedologic, and hydrologic flow systems, for instance, principles of gradient, resistance, and efficiency selection operate to differentially favor steeper flux gradients, lower resistance (preferentially preserving higher resistance materials and structures), and more efficient network patterns. This makes certain patterns and features more probable and reinforces and preserves them once developed. Examples include development of fluvial channels and channel networks, karst conduits, and soil water preferential flow paths.

Selected pathways also occur in the coevolution of vegetation and hydrological systems, in the sense of optimizing or maximizing water fluxes. Mikhail Budyko pioneered this idea in the 1970s, and others, such as Peter Eagleson, refined it and more explicitly tied it to selection principles. For example, plant cover often develops via selection to maximize water fluxes (if energy, nutrients, or other resources are not limited). Where faster or more efficient biogeochemical transformations and cycles confer some survival, reproductive, or other advantage, selection can also promote pathways toward maximum biogeochemical cycling rates and primary productivity.[7]

Now we circle back to the question raised earlier. Given the immense geographical variability of environmental factors and controls at any one place, and their equally immense variability in time, coupled with dynamical instability and historical contingency, why aren't there *even more* different landscape evolution trajectories observed? Well, because some trajectories that are plausibly imaginable turn out to be impossible on closer inspection. And some that were possible in the past are no longer possible because the conditions that made them possible no longer exist. By extension, of course, some trajectories happening or possible now may become impossible in the future. Still others are possible, but not probable, because they require uncommon events or boundary conditions. Still other trajectories occur, or may occur, but are rarely observed because they are transient.

That leaves us in our typology with non-repeatable and repeatable recurring paths, many of which are favored by various kinds of selection. These selection principles account for the fact that some pathways, or types thereof, are far more probable and therefore more frequently observed.

HISTORIES, OBSERVED AND INFERRED

The perspective on evolutionary trajectories is fundamentally different looking back—historical studies, environmental reconstructions, or interpreting contemporary landscapes—versus looking forward. For an individual landscape at a particular time, only one historical sequence or chain of events led to the observed landscape. That does not necessarily mean that only one history could have produced the observed landscape, though that is probably true in many cases. It also does not reflect that at any point in the past other trajectories or variations could have happened, although only one did.

In predicting future evolution, however, the number of potential pathways may be large. Thinking in terms of the types of pathway described above yields some simple axioms as a starting point when looking back or ahead:

1. From a historical perspective, only one sequence of events led to an observed landscape and is potentially reflected in it. Multiple pathways were possible, but only one sequence of events occurred. However, when the actual pathway is unknown, multiple possibilities must be considered.
2. With respect to prediction, there are no observed pathways and more than one—perhaps many—possible pathways. However, the number of potential pathways, though potentially large, is finite.
3. The number of possible future pathways can be constrained by identifying impossible and improbable ones and eliminating them from consideration. Principles of canalization based on current conditions usually allow further constraining.
4. Evidence of transient pathways may not be, and often will not be, preserved in observed landscapes.
5. The number of possible past pathways leading to an observed landscape can be narrowed down by identifying impossible and improbable ones and eliminating them from consideration. Benchmarks (historical evidence or observations of landscape states) can allow further refinement of the possibilities.
6. There must exist at least as many observed pathways as there are recurring pathways.

MATURATION

Landscape evolution is ongoing, and identification and analysis of evolutionary trajectories requires enough time for patterns to emerge. In *Landscape Evolution*, I described this phenomenon using graph theory formalisms, but the idea is that an evolutionary sequence can be called mature when there have been enough transitions for the historical pattern (or graph structure) to emerge.

Because evolution is ongoing, the sequences represented by chronologies, chronosequences (see below), and evolutionary graphs develop over time. The appearance of each state or form is a historical transition. An evolutionary sequence may be called *mature* when enough transitions have occurred for the graph structure to become evident. An ecological STM for rangelands is mature, for instance, when all plausible ecological states (or at least all those relevant to the management context or research question) have been identified. A phylogenetic tree (a graphical representation of evolutionary relationships among organisms), for another example, has matured when enough nodes and branches have been identified to define its branching structure.

How do we know when maturity is reached? There are no firm rules or criteria. Some general guidelines are that more time has passed than the general time frame of analysis or that more (potential) transition trigger events have occurred than there are observed transitions.[8]

DIVERGENCE AND CONVERGENCE

The average magnitude of spatial variability in landscapes—for example, elevation contrasts, differences in vegetation coverage or microbial communities, variations in soil properties—can do three things as landscapes evolve: remain constant, increase, or decrease. *Convergent* evolution is marked by decreasing variability—e.g., more uniform plant cover or reduction of topo-graphic relief (as when higher spots are worn down and lower ones infilled). During convergence, the landscape is moving to-ward more spatial uniformity. Increasing spatial variability, such as patchier vegetation cover or increasing relief, is the hallmark of *divergent* evolution. During divergence, there occurs, on aver-age, increasing difference between locations within the landscape. What if there is little or no change in spatial variability either way? This could occur in a static landscape (possible over small areas and short periods of time but unlikely otherwise) or where divergent and convergent trends offset each other and cancel out—I mention the latter as a theoretical possibility, but I cannot think of a good real-world example. Different aspects of the same landscape can undergo convergence and divergence—for instance, soil thickness could become increasingly variable while cation exchange capac-ity converges toward more uniformity—but these do not normally directly offset each other.

In biological evolution, convergence occurs when adaptations in different places that have similar selection pressures converge on similar forms or, more generally, when similar evolutionary adaptations occur for any reason other than the organisms having common ancestry. For example, the C4 photosynthetic pathway (one of three known ways for plants to fix carbon) in plants and flight in fauna (birds, insects, bats) have evolved repeatedly along different branches of the tree of life. Convergent landscape evolution is a broader concept than convergence in biological evo-lution and includes, for instance, *ecological equivalents*, where different species occupy analogous niches and perform similar roles in different locations. For example, multiple different grasses, sedges, and rushes occupy the primary producer role in freshwater marshes.

While biological evolutionary divergence is far more likely to be driven by divergent landscape evolution than to drive it, bio-logical evolutionary divergence (whereby related species evolve differing traits) can sometimes lead to increased spatial variability of landscapes. Niche construction and changes in ecosystems can be stimulated by evolutionary change arising from genetically based interactions. This can create or reinforce spatial variations in landscapes. This has been most thoroughly studied in plant-soil interactions. For instance, different tree genotypes have different microbial communities in their rhizospheres. The microbes affect soil nutrients, and those in turn influence plant performance, as described in Chapter 2.

Convergent vs. divergent evolution is the single most important manifestation of dynamical stability or instability in landscapes. Stable systems can return to or toward the preexisting state after a change or disturbance (as long as the latter is not too large or severe). Dynamically stable systems are insensitive to variations in initial conditions and on average tend to smooth out initial variations. Dynamical stability is therefore linked to convergence (or steady state, with no net change in variability). Dynamical instability is associated with divergent evolution. Unstable systems, by definition, are subject to disproportionally large and long-lived deviations from preexisting states after changes or perturbations. They are also sensitive to variations in initial conditions, which tend to be exaggerated over time.

Divergence and convergence are both common, though my own research has focused on divergence, and traditional ecological and geoscience research has emphasized convergence. You can discern this case by case in the scientific literature, but it has been directly addressed in some areas of landscape science. In ecosystem development, trends toward increasing homogenization are common (as, for example, when successful colonizing or introduced species take over a landscape). So are trends that show increasing heterogeneity and patchiness of factors such as soil chemistry, vegetation composition and cover, and microbial communities. While the interpretations have been hotly debated at times, the presence of both convergent, relief-decreasing, and divergent, relief-increasing topographic evolution has long been recognized. Divergence and convergence in pedogenesis (by the tongue-twisting names of *proanisotropic* and *proisotropic pathways*) has been explained by the likes of Francis Hole (1913–2002) back in 1961 and Donald Johnson in the 1980s.

At this point, you will not be surprised to learn that divergent evolution of landforms, soils, and biological communities often occurs in response to the same drivers and controls and in simultaneous, mutually adjusting ways. Pavel Šamonil and coworkers in the Czech Republic, for instance, have shown how tree uprooting by windthrow triggers local divergence in geomorphological, soil, and ecological processes. Their work also shows, however, that intense podzolization can overprint local variability and divergence, producing convergent pedogenesis.[9] Several studies have found convergence of soil types at broad spatial scales, coupled with divergence at more local (hillslope or field) scales within the same landscape.

In landscapes, nothing is forever, and that goes for convergence and divergence. Neither can continue indefinitely. Thresholds or fundamental limits (e.g., base level, maximum photosynthetic efficiency, finite supplies of matter and energy, and rates of chemical kinetics) ultimately limit divergent development, resulting in a mode switch from divergence to convergence. Convergent pathways are also limited, though this is more difficult to prove theoretically. However, empirical circumstantial evidence is ample in the form of the absence of landscapes that are uniform, even in very old, geologically stable settings.

CHRONOSEQUENCES

A key tool in reconstructing landscape evolution is the *chronosequence*. Because landscape evolution typically proceeds too slowly for direct observation, we must use other approaches. A chronosequence is a space-for-time substitution. Envision, for instance, a glacier that has been steadily retreating for one hundred years. The ground exposed between where the ice was a century ago and where it is now represents a hundred-year history of ecological succession and soil development. The basic idea (assumption) of a chronosequence is that all important environmental factors except time or age are constant.

Take, for instance, a river valley with a step-like sequence of alluvial terraces (Figure 4.8). Moving from the river to the active floodplain upward and outward up the terrace staircase, you also move from younger to older surfaces. If similarity of the alluvial deposits and equivalent environmental conditions throughout the time of the sequence can be assumed, these surfaces differ principally in their age. Comparing weathering, topography, soils, vegetation, or other properties along the terrace staircase is then a proxy for development over time. Chronosequences have been widely used in pedology, ecology, and geomorphology.

Figure 4.8. The flat surfaces are alluvial terraces formed when a period of limited river incision and floodplain formation is interrupted by an episode of downcutting (triggered in this landscape on the South Island of New Zealand by tectonics and climate change), stranding the former floodplain above the river. The terraces thus represent a chronosequence from the youngest, lowest, and closest to the modern river to the oldest, highest, and farthest away.

Criticizing the chronosequence approach is not difficult. Principally, true equivalence of all factors except time or age is both
rare and difficult to prove or even assess. In the alluvial terrace example, the surfaces also differ in elevation—and therefore likely in contemporary drainage and hydrology. The former
floodplains will also have undergone different histories of climate,
land use, and disturbances. But where detailed historical reconstructions at a single location are not feasible—as is often the
case—space-for-time methods may be the only option. The utility of chronosequences, notwithstanding their flaws, has not been
seriously challenged in pedology and geomorphology, even as
the detailed reconstructions and interpretations may be fiercely
debated. The general, albeit qualified, acceptance of chronosequences is not necessarily the case in ecology, but many ecologists
accept the chronosequence approach, while acknowledging its
flaws, when no better alternative is available (which is, again,
often the case). Chronosequences can be frustrating if one is
searching for clear, monotonic developmental trends, which are
often not evident. But devaluation or rejection of chronosequences
because they fail to show clear trends is not valid, or even relevant, when the goal is to identify trajectories of environmental
change.[10] Emerging studies of succession on ground exposed
by glacier retreat within the past century or so indicate that local environmental heterogeneity, sometimes reinforced by the
biota, exerts more control over vegetation and microbial community composition that distance from the glacier edge (a surrogate
for time).[11]

Chronosequence studies have often used single or mean char-
acterizations of the stage or state—that is, each member (terrace
surface, abandoned field, etc.) is considered to have a characteristic soil, microbial community, degree of erosional dissection,
or what have you. But chronosequences also have spatial variation within as well as between members. Different states (e.g.,
soil types) can exist during the same time and therefore on the
same chronosequence member. This need not be a drawback,

as comparison of spatial variations along the sequence can directly address the possibility of divergent or convergent evolution. Chronosequence studies that explicitly addressed variations within as well as between members and changes in variability or diversity have mostly (but not always) revealed an increase in soil richness over time. This indicates pathways more complex than a linear sequence of progressive pedogenesis. Some studies have shown distinctly different pedogenetic pathways within a single chronosequence. Multiple pathways and stable states identified via ecological chronosequences have led to several alternative frameworks such as community assembly, alternative stable-state models, and STMs. As mentioned above, divergence in chronosequences can be exploited to explore divergent ecosystem landscape evolution. STMs are in essence a type of chronosequence developed using evidence other than, or in addition to, space-for-time substitutions.[12]

ROBUSTNESS AND PATH STABILITY

Chronologies, chronosequences, and STMs represent (what we know of) the historically contingent evolutionary pathways of landscapes. To what extent are the sequences potentially repeatable vs. idiosyncratic? This is linked to path stability or robustness of these pathways. We are not talking here about the stability of a particular state of the landscape, such as a boreal forest on podzolized soils or an intertidal mangrove swamp. We are talking about the stability of historical sequences themselves, such as those that the forest or swamp might be a part of.

Path stability extends stability concepts to the domain of networks of state changes and necessitates a brief side trip into stability concepts and terminology. These include *mechanical stability* or similar notions of resistance, including resistance to change or proximity to critical thresholds. Examples include evaluations of hillslope stability with respect to slope failures, engineering design of stable channels, and ecological resistance to disturbance by, for example, drought. A second concept, *dynamical stability*, is used extensively in this book. This deals with the ability of an ESS to recover or return toward the previous state following a change or disturbance. Dynamical instability is manifested as sensitivity to minor variations in initial conditions and small perturbations and is thus tantamount to deterministic chaos. The tendency or ability to maintain a developmental pathway or mode of development despite environmental fluctuations is assessed by *path stability*. This complements stability concepts linked to resistance and resilience with an assessment of *robustness*. Path stability applies to a network of changes in system states—such as that represented by an STM or chronology.

The method for assessing path stability is based on representing a historical sequence as a graph or network. System states or developmental stages are the nodes, and transitions between them are the links connecting the nodes. The simplest examples are one-way, irreversible successional sequences. More complex possibilities occur when transitions are reversible (e.g., due to disturbances) and where development bifurcates (or *n*-furcates) so that several landscape states may occur at a given stage or age.[13]

I first applied this to a regional soil STM-type chronosequence for sandy uplands of the outer coastal plain of North Carolina (Figure 4.9), including seven categories of soil taxa and twenty-three links between them. This includes positive facilitation links, where the existence of a soil type enables transition to another, and negative inhibition links, whereby once a particular soil type is formed, transition to some others is prevented or impeded. A vegetation succession model for the same region as well as some archetypal chronosequence structures were also analyzed. These analyses showed that sequential pathways where reversal or retrogression is impossible or inhibited are path stable. So too are divergent *or* convergent patterns (from a single starting point, development diverges toward multiple end points connected only to the original state or multiple unconnected starting states converge to a single end state). Somewhat less robust are reversible linear progressions—a sequential path where reversal is *not* inhibited. This could occur, for example, where a linear progressive sequence occurs that can be reset or reversed by disturbance. Some divergent and convergent networks also fall into this category.

Less robust still are low-divergence paths, where disturbances to the pathway result in divergent evolution, but the rate and magnitude of divergence are relatively low. With respect to archetypes, this applies to a sequence where progression is promoted (each stage promotes transition to the next), regression or reversal is inhibited, the final state is self-maintaining, and intermediate states have no self-effects (they are not self-maintaining or self-limiting). The soil chronosequence shown in Figure 4.9 falls into this category. Somewhat less path-stable chronosequences occur when the final state is a "sink" state that is difficult to dislodge, working against repeatability of the sequence. The least robust and path-stable sequences are termed *complex high-divergence*. Networks with a random pattern of stage or state transitions, or where transitions between any two states are possible, fall into this category.

I also applied this kind of analysis to evolution of fluviokarst landscapes in central Kentucky, considering historical transition networks separately for landforms, soils, and hydrological flux patterns, and to a generalized model of hydrological flow partitioning applicable to many landscapes. Path stability was evaluated under different assumptions and scenarios. The upshot is that results so far suggest that path stability and robustness in landscape evolution are rare and that complex high-divergence patterns, which are path-unstable and non-robust, are common.[14]

Path-stable landscape state-transition patterns have reliable memories in that in knowing the current state, you can infer the previous state. For the highest path-instability pattern, where any transition is possible, a given landscape state provides no information on what the previous state might have been.

Figure 4.9. Examples from a chronosquence from the North Carolina coastal plain. Top is a minimally developed Quartzipsamment on a barrier island dune. In the middle is a Hapludult and at the bottom a Paleudult, both formed from sandy barrier island parent materials. The Psamments are readily transformed into Hapludults as clay accumulates in the subsoil. Continued progressive development forms a Paleudult. Because the argillic horizons are resistant, regressive development from Hapludult to Psamment or Paleudult to Hapludult is rare. These transitions are only part of a complex sequence better characterized as a state-and-transition model than a successional sequence.

See references in endnote 13.

RAVINE SWAMP MEMORIES

To illustrate some of these ideas, let's wade into the swamp—specifically, a ravine swamp near the Neuse River estuary in North Carolina, like the one shown in Figure 4.10.

Figure 4.10. Ravine swamp.

When sea levels were lower, earlier in the Quaternary, tributaries to the lower Neuse River had greater slopes and incised steep (well, by coastal plain standards) valleys. As sea levels rose during the Holocene, the river and its valley were submerged, and the tributary valleys truncated. Now, these typically terminate at the estuarine shoreline in a hardwood swamp. Because of their unusually steep valley walls (again, by local and regional coastal plain standards), I call these *ravine swamps*. They are ponded and inundated year round and dominated by water tupelo (*Nyssa aquatica*) and bald cypress *(Taxodium distichum)* trees, along with a smattering of other trees typical of bottomland hardwood swamps of the Carolina coastal plain. Though they are occasionally overwashed by estuary waters during large storms, they are freshwater. Some discharge to the estuary via a small stream; in other cases, the outlet is a dripline. They often do not discharge during dry spells, but the terminal swamps never dry out.

The steep tributary valleys are an inherited feature from previously lower sea levels; these drainages are not actively incising now due to their lower channel slope gradients (Figure 4.11). Another important inherited feature lies below the ravine swamps: a swamp paleosol (Figure 4.3). The shoreline bluffs surrounding the ravine swamps are mainly composed of the Pleistocene Flanner Beach formation, a roughly 200 ka old sequence that records a previous sea-level transgression from estuarine facies to nearshore and barrier island sands. The swamp paleosol at the base is the upper part of the James City formation, making it 700 to 1,000 ka. This paleosol, often exposed on eroding shorelines next to the ravine swamps, contains numerous tree trunks and roots, mainly bald cypress. This clayey, organic-rich, root-reinforced material is much more resistant to erosion than the Flanner Beach sediments and has very low permeability. Therefore, this inherited feature strongly affects the modern ravine swamps by leaving them perched roughly a meter above contemporary sea and estuary level. The tributaries have insufficient stream power to cut through the swamp paleosol (larger tributaries with more force have done so), and not enough water can percolate through it to move as groundwater.

The tupelo and cypress trees standing in the water are a bit of a mystery. Both species, once established, can tolerate constant inundation and are indeed iconic features of swamps of the southeastern United States. But neither species can germinate and establish itself in standing water. Once established, cypress and tupelo can reproduce via stump sprouting (tupelo is much better at this than cypress) or by fallen trunks or stumps serving as "nurse logs" for seeds of either species, which are dispersed by floating in water. Rootwads of trees uprooted by storm winds can also serve as above-the-waterline sites for new tree establishment. But the mystery is how the trees got there in the first place.

The origin of the tupelo-cypress stands in the ponded areas—again, they have never been known to dry out—cannot be explained based on current or recent environmental conditions.

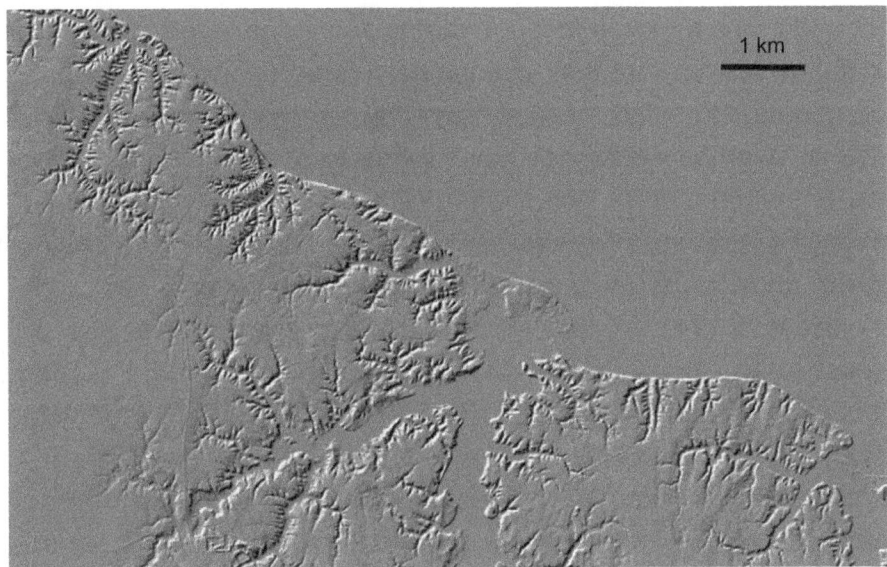

Figure 4.11. Shaded relief map showing the dissected valley side slopes of the Neuse River estuary near Havelock, North Carolina.

So, some sort of historical explanation is necessary. Let's walk through some of the possibilities, using some of the ideas from earlier in this chapter.

First, let us postulate that the ravine swamps were colonized by tree species that *can* germinate in standing water, and that these served as nursery sites for subsequent plants. We can lump this potential pathway into the *impossible* category. No such woody plant is known in the region—while many can germinate in some very wet mud, all require at least brief periods of subaerial exposure. The stumps and roots in the swamp paleosol are cypress, so it seems earlier swamp trees were of the same species, or at least genus, as the modern ones.

A second possibility is that the swamps once dried out seasonally or during droughts, allowing seedlings to gain a roothold. This is how tupelo gum and cypress often establish on stream banks, where seeds take hold during low-water periods and are viable enough when the water comes back up to survive. This explanation requires that a past climate or hydrologic regime featured dry spells that allowed the swamp vegetation to get started and/or that they formed in their current location during lower sea levels, with lower water tables, so that they occasionally dried enough for seeds to germinate. Both fall into the *currently impossible* category—the ravine swamps could not have formed under recorded historical or contemporary conditions. With respect to whether these might have been the case earlier in the Quaternary, I consider them *improbable* pathways. No paleoclimate records indicate a significantly drier climate or record the presence of a true dry season, and the fact that they all exist only near the modern tributary mouths suggests formation in the more recent geologic past.

In 2018, Hurricane Florence deposited 60 to 100 centimeters of sand in portions of the ravine swamps, transforming those parts of them from permanently flooded ponded swamps with a mucky clay substrate to only occasionally flooded areas with a sandy substrate. This kind of event could certainly get the trees going— new cypress seedlings could be seen in the storm deposits within a year. Augering into the ravine swamp sediments did not reveal any previous similar storm deposits—though I have only ever been able to get a couple of meters down with an auger. Deeper coring might tell a different story. Is it also possible that enough sediment was deposited in a past storm to allow tree establishment and subsequently flushed out by tributary flows? If so, it would qualify as a *transient* pathway, as no evidence I can see has been left. I am skeptical of this mechanism due to the low transport and erosive capacity of the tributaries, but it must be left on the table.

Another impact of Hurricane Florence was to deposit a great deal of large woody debris (that's scientist for logs) in some of the tributary valleys. As these logs settle and decompose, they will provide a suitable substrate for establishment of various trees. In a setting where they eventually sink and rot into, or are covered by, standing water, tupelo and cypress would be the main survivors and could maintain their populations. This is my favorite explanation at the moment.

So, what do the ravine swamps "remember?" They recall sea-level changes, reflected in the tributary and valley morphology. Once the tributary valley began incising, positive feedbacks and canalization ensured that the channel would grow and persist. They remember a Pleistocene swamp, currently influencing the resistance and hydrology of the modern swamp. They recollect some past condition or disturbance—perhaps a storm-load of large woody debris—that allowed *Nyssa aquatica* and *Taxodium distichum* to become established. Some canalization of forest composition ensues in the flooded area, as no other tree can establish in the pond, and the *Nyssa* and *Taxodium* can maintain their populations.

Hurricane Florence represents a conditionality switch, at least in the short term and probably longer. As outlined in a 2022 article that included assessment of the storm's impact on the ravine swamps, in addition to place characteristics of the environment of and near the ravine swamps, the particulars of the storm track and climatology were critical for the impacts on the swamps. Portions have been ecologically, geomorphologically, and pedologically transformed.[15]

SUMMARY: HISTORY AND MEMORY IN LANDSCAPE EVOLUTION

History matters.

We need laws and global principles to explain landscape evolution. But laws and global principles alone are insufficient. Accounting for environmental and geographical context—place factors—is also necessary but insufficient. History also must be incorporated to explain landscape evolution (which, after all, has temporal dimensions by definition). Inheritance and legacy effects, disturbance records, temporal changes in driving factors such as climate, tectonics, and human impacts, stages of development, and path dependence—all must be incorporated.

Landscape evolution is historically contingent.

Types of historical contingency include inheritance and legacy effects and conditionality. Bailey effects, where small actions, disturbances, or changes may have large and increasing influences as history unfolds, also occur. Dynamical instability helps explain both Bailey effects and the sometimes disproportionately large growth over time of variations in initial conditions. Canalization is an important form of historical contingency, relevant to situations whereby once an evolutionary path has been established, other previously possible pathways are eliminated, and the trajectory is channeled, or canalized. Canalization is a key mechanism for locking in effects of small changes or inherited features.

Multiple possible pathways and outcomes are typical in landscape evolution.

Linear sequential models may accurately reflect some specific historical sequences. But over longer time periods and/or multiple landscapes, more complex patterns often occur. These are better described by STMs (*sensu lato*) that can incorporate multiple pathways, cycles, and reversals. A historical sequence or network of transitions is considered mature when it has become large or lengthy enough to confidently identify its basic structure.

Potential landscape evolution paths may be reinforced or inhibited.

Why does the vast range of possibilities (given multiple-path development in Earth surface systems, the critical role of disturbances, constant environmental change, and the prevalence of dynamical instabilities and thresholds) produce a relatively smaller number of observed patterns? Why do some phenomena occur repeatedly? The selection for, and reinforcement of, some pathways and the prevention or inhibition of others account for this.

Impossible evolutionary trajectories are prohibited by general laws or are dependent on boundary conditions that do not exist. This is not necessarily a statement of the obvious; our comprehension of the possible is mutable, as understanding of general laws and boundary conditions increases. Currently impossible pathways have, or could have, occurred. But they are dependent on factors that no longer exist, such as past climates or extinct organisms. Possible but rare are the improbable and inhibited evolutionary pathways, which are associated with rare events or conditions, or resistance factors and counteracting feedback mechanisms, respectively. Transient paths in landscape evolution also exist, though this is a highly scale-contingent concept. Transient trajectories may be frequent or rare. Either way, they are rarely observed because they do not last long and/or are not well preserved. Pathways that occur in different locations but cannot recur at a specific site due to self-limiting or irreversible processes are recurring but non-repeating trajectories. Recurring pathways may be repeated at the same location due to occasional rejuvenation or cyclical inputs or boundary conditions. Recurring pathways that are frequently observed are *selected*. Selection principles increase the likelihood that these pathways will occur at multiple times and locations.

Landscape evolution may be convergent or divergent.

Three things can happen with respect to spatial variability in landscapes as they evolve. Variability can decrease, remain constant, or increase. Decreasing anisotropy is convergent, in that the landscape or some property converges toward more spatial uniformity. Divergence is associated with increasing spatial unevenness as locations within the landscape, on average, become increasingly different. Neither divergence nor convergence can continue indefinitely, as thresholds or fundamental limits are reached. "Remaining constant" rarely lasts long, as change, disturbance, or internal instabilities or thresholds are likely to tip the landscape into convergent or divergent modes.

Historical sequences may be stable or unstable.

Path stability or robustness reflects whether a sequence is likely to be repeated and the extent to which it may diverge over time. Historical sequences represented as graphs or networks can be analyzed to determine whether they are path stable, reversible progressions, low divergence, or high divergence. Path stability seems to be rare in landscape evolution, whereas complex, high divergence is common. Path stability analyses point to the importance of a sink states—self-perpetuating states that, once entered, there is no way out of (without major changes or disturbances). Path stability investigations also highlight difficulties of generalizing regarding landscapes without reference to local historical and geographic contingencies.

NOTES

1. V. O. Targulian and S. V. Goryachkin, "Soil Memory: Types of Record, Carriers, Hierarchy and Diversity," *Revista Mexicana de Ciencias Geologicas* 21 (2004): 1–8; V. O. Targulian and S. V. Goryachkin, "Soil Memory and Environmental Reconstructions," *Eurasian Soil Science*, 44 (2011): 464–465; V. O. Targulian and M. A. Bronnikova, "Soil Memory: Theoretical Basics of the Concept, Its Current State, and Prospects for Development," *Eurasian Soil Science* 52 (2019): 229–243.

2. On general concepts of legacies and memories: M. E. Swanson et al., "The Forgotten Stage of Forest Succession: Early-Successional Ecosystems on Forest Sites," *Frontiers in Ecology and the Environment* 9 (2011): 117–125; N. Ferreiro, P. Satti, and M. J. Mazzarino, "Biological Legacies Promote Succession and Soil Development on Tephra from the Puyehue-Cordon Caulle Eruption (2011)," *Austral Ecology* 43 (2018): 435–446; L. R. Walker et al., "Biological Legacies: Direct Early Ecosystem Recovery and Food Web Reorganization after a Volcanic Eruption in Alaska," *Ecoscience* 20 (2013): 240–251. On pyrophytic vegetation: M. Blackhall et al., "Effects of Biological Legacies and Herbivory on Fuels and Flammability Traits: A Long-Term Experimental Study of Alternative Stable States," *Journal of Ecology* 105 (2017): 1309–1322. On memory in paleoecology: B. M. Benito, G. Gil-Romera, H. J. B. Birks, "Ecological Memory at Millennial Time-Scales: The Importance of Data Constraints, Species Longevity and Niche Features," *Ecography*, 43 (2020): 1–10.

3. E. Lekevicius, *The Origin of Ecosystems by Means of Natural Selection* (Vilnius: Lithuanian Academy of Sciences, Institute of Ecology, 2002).

4. S. N. Lane and K. S. Richards, "Linking River Channel Form and Process: Time, Space and Causality Revisited," *Earth Surface Processes and Landforms* 22 (1997): 249–260; K. Beven, "What We See Now: Event-Persistence and the Predictability of Hydro-Eco-Geomorphological Systems," *Ecological Modelling* 298 (2015): 4–15. K. C. Parker, "Effects

of Complex Geomorphic History on Soil and Vegetation Patterns on Arid Alluvial Fans," *Journal of Arid Environments* 30 (1995): 19–39.

5.   H. A. Gleason, "The Individualistic Concept of the Plant Association," *American Midland Naturalist* 21 (1939): 92–110; W. H. Renwick, "Equilibrium, Disequilibrium, and Nonequilibrium Landforms in the Landscape," *Geomorphology* 5 (1992): 265–276. See also K. Beven, "A Manifesto for the Equifinality Thesis," *Journal of Hydrology* 320 (2006): 18-36.

6.   I originally presented these ideas in: J. D. Phillips, "Evolutionary Pathways in Soil-Geomorphic Systems," *Soil Science* 184 (2019): 1–12.

7.   This example is revisited in more detail in Chapter 5.

8.   For instance, for post-fire vegetation transitions, a mature evolutionary sequence may be considered to have occurred when sufficient time has elapsed for succession to occur and more post-fire recoveries have been observed than there are observed transitions.

9.   P. Samonil et al., "Converse Pathways of Soil Evolution Caused by Tree Uprooting: A Synthesis from Three Regions with Varying Soil Formation Processes," *Catena* 161 (2017): 122–136; P. Šamonil et al., "Impacts of Old, Comparatively Stable, Treethrow Microtopography on Soils and Forest Dynamics in the Northern Hardwoods of Michigan, USA," *Catena* 140 (2016): 55–65; P. Šamonil et al., "Soil Mixing and Genesis as Affected by Tree Uprooting in Three Temperate Forests," *European Journal of Soil Science* 66 (2015): 589–603.

10.   Examples of debate over and critical analysis of the chronosequence approach include B. L. Foster and D. Tilman, "Dynamic and Static Views of Succession: Testing the Descriptive Power of the Chronosequence Approach," *Plant Ecology* 146 (2000): 1–10; R. J. Huggett, "Soil Chronosequences, Soil Development, and Soil Evolution: A Critical Review," *Catena* 32 (1998): 155–172; E. A. Johnson and K. Miyanishi, "Testing the Assumptions of Chronosequences in Succession," *Ecology Letters* 11 (2008): 419–431; A. D. M. Paine, "Ergodic Reasoning in Geomorphology," *Progress in Physical Geography* 9 (1985): 1–15; D. Sauer, "Approaches to Quantify Progressive Soil Development with Time in Mediterranean Climate-I. Use of Field Criteria," *Journal of Plant Nutrition and Soil Science* 173 (2010): 822–842; R. J. Schaetzl, L. R. Barrett, and J. A. Winkler, "Choosing Models for Soil Chronofunctions and Fitting Them to Data," *European Journal of Soil Science* 45 (1994): 219–232; L. R. Walker et al., "The Use of Chronosequences in Studies of Ecological Succession and Soil Development," *Journal of Ecology* 98 (2010): 725–736.

11.   See, e.g., J. Brandani et al.,. "Homogeneous Environmental Selection Structures the Bacterial Communities of Benthic Biofilms in Proglacial Floodplain Streams," *Applied and Environmental Microbiology* 89 (2023): 10.1128/aem02010-22; L. Sieffried, P. Vittoz, and S. N. Lane,. "Hydrological Heterogeneity and the Plant Colonization of Recently Deglaciated Terrain," *Arctic, Antarctic, and Alpine Research* 55 (2023): 2259677.

12.   Many methods for developing chronologies of landscape development exist and are briefly summarized in Chapter 3 of *Landscape Evolution,*

and these are outlined in recent textbooks and monographs in paleoclimatology, paleopedology, paleoecology, and Quaternary geology. I say
recent because technological and methodological innovations in dating and
historical inference are ongoing.

13. Mathematical details are given in Chapter 5 of *Landscape Evolution*, and in
     J. D. Phillips,. "The Robustness of Chronosequences, *Ecological Modelling*
     298 (2015): 16–23.

14. Details of the robustness analysis are given in Chapter 5 of *Landscape
     Evolution*, which is in turn based on data from these papers: J. D. Phillips,
     "Landform Transitions in a Fluviokarst Landscape," *Zeitschrift Fur Geo-
     morphologie* 61 (2017): 109–122; J. D. Phillips, "Historical Contingency in
     Fluviokarst Landscape Evolution," *Geomorphology* 303 (2018): 41–52.

15. J. D. Phillips, "Geomorphic Impacts of Hurricane Florence on the Lower
     Neuse River: Portents and Particulars," *Geomorphology* 397 (2022):
     108,026.

5

The Goals (If Any) of Landscape Evolution

INTRODUCTION

What are landscapes trying to accomplish as they evolve?

Perhaps nothing. While one might ascribe goals and objectives to individuals or to social organizations of organisms, flowing water certainly has no plans or aspirations. Geochemical reactions and heat fluxes are aimless. Gravity and pressure gradients do what they do with no concern for the outcomes; rocks, sediments, and air masses care not what happens to them. Landscapes lack ambition.

Yet, it is natural to wonder where evolution is headed. Our senses and intuition (whether or not we believe in some sort of guiding hand) cannot accept that it is all pointless or random; if not an endpoint or destination, there must be some pathways and states that are more ideal than others—some possibility of improvement in some sense or another.

Landscape scientists have indeed picked up on trends and patterns in the development of landforms, topography, soils, ecosystems, and hydrologic systems. We have noted that some forms, patterns, structures, and developmental trends work better than others by certain criteria and seem to crop up repeatedly. And in many cases, it appears that in some apparently mysterious ways landscape evolution is guided toward certain outcomes.

Mysterious Ways. Jonathan D. Phillips, Oxford University Press. © Oxford University Press (2025).
DOI: 10.1093/9780197755129.003.0005

Goal function is an umbrella term for hypotheses that Earth surface systems (ESS) progress toward specific conditions that minimize, maximize, or optimize something, or toward more stable or persistent states, or even toward some final, ultimate destination. We'll review some of these in this chapter. Goal functions include several different criteria that have been proposed to govern the operation and development of landscape systems.

Attractor is a broader term referring to a state toward which a system evolves, with no necessary connotations about why it does so. In mathematical dynamical systems theory, an attractor is a set of numerical values toward which a system tends to evolve, for a wide range of starting values. Climax vegetation communities, mature climax soils, peneplains, steady-state equilibria, and self-organized criticality are all examples of attractors that have been proposed for landscape systems (Chapter 1).

BOWLING FOR ATTRACTORS

Let's start with a metaphor: the game of bowling. The ball may roll down the lane along different trajectories but within a narrow, well-defined range. Different scoring outcomes occur, but again within a fixed, clearly defined range. No matter what, the ball ends up in the same place, the depression behind the pins. The bowling metaphor applies to single-path, single-outcome notions such as vegetation succession, erosion to a peneplain, progressive development toward a climax soil, and (single) steady-state equilibrium or self-organized critical attractors.

The latter are bowling-type situations, in that there is some variation in pathways, but the range is relatively small, the general direction is constant and predetermined, and the ultimate outcome is preordained. If you could temporarily stop the bowling ball, or roll it back up the lane, that would be the analogy to how disturbances work in deterministic, single-outcome concepts. Where the metaphor falls short is that in any bowling lane the outcome is the same, whereas the ultimate state of landscapes is generally conceived to vary in different environments. These concepts fall into three general categories: succession-to-climax, equilibrium, and relaxation-time.

SUCCESSION TO CLIMAX

Ecological succession toward a climax community, progressive soil evolution toward a mature zonal (or climax) soil, and the geomorphic cycle of erosion through youthful, mature, and old age to a peneplain are three examples of succession-to-climax, literally or figuratively. These are reviewed in Chapter 1 and are associated with convergent evolution as described in Chapter 4.

Ecological succession models typically stipulate a stable equilibrium, self-maintaining climax community determined chiefly by climate. These days, deterministic succession to a preordained climax attractor is considered by many ecosystem scientists to be a particular case from among a larger range of possibilities. Nonetheless, the notion of a climax community as an intrinsic geographical property—i.e., the climax vegetation is a characteristic of a site, along with its geology or climate—remains ingrained in natural resource management and inventory, vegetation and ecosystem mapping, ecoregion delineation and mapping, and modeling ecological responses to climate. Note that these applications do not require or necessarily imply the acceptance of traditional climax theory. But their use does reflect persistence of the general idea of climax communities as a normative state or geographical property, even though it may be tacitly recognized as a simplifying assumption.

In soil mapping and pedology, there is no logical necessity tying the state factor approach to any single concept of soil evolution. However, the factorial approach has historically been linked to single-path, single-outcome progression to climax or zonal soils. Dokuchaev's seminal work in the 1880s included the concept of climatically determined climax soils, apparently arising from his work in the Russian steppes, where climatic zonation of soils is perhaps better expressed than anywhere else. The factorial approach was also embraced by Curtis Marbut (1863–1935) in the 1930s, who argued that, if undisturbed, most soils will evolve to a zonal, climax form. Marbut, who was chief of US Soil Survey operations, called these *normal soils*, further cementing the idea of a normative status for the climax forms. US soil taxonomy was codified about this time based on three major categories: zonal, azonal, and intrazonal. Zonal soils are normal, climax soils. Azonal are young, immaturely developed soils presumably developing toward their zonal climax, and intrazonal is essentially an "other" class. Intrazonal soils in this scheme were seen to be dominated by local or idiosyncratic factors of soil formation. The evolutionary pathway presumably runs from parent material to azonal soils to zonal soils, with whatever doesn't fit lumped into the intrazonal category.

In progressive pedogenesis toward zonal, climax soils and in ecological succession, it was generally recognized that disturbance could reset the pathway (though the view of disturbance changed from conceiving disturbances as aberrations to recognizing them as natural and inevitable in many settings). By contrast, William Morris Davis's late-nineteenth and early-twentieth-century theories of geomorphic landscape evolution are explicitly cyclical. The progression of denudation through youth, maturity, and old age stages is restarted by uplift to reset the cycle. Between these uplift events, the cycle of erosion is one-way, involving progressive lowering of mean elevations and decline of slope gradients until the "climax" is reached—a low relief peneplain graded to base level. Rejuvenation of denudation by uplift certainly happens, but there is scant evidence to support a pulse of uplift followed by a long period with no uplift as a common sequence. By contrast, landscape disturbances that reset the clock for ecosystem and soil development are widely documented.

Climax communities and zonal soils both recognize that the climax forms vary geographically (governed mainly by climate and geology). The implicit assumption is that climax forms are adjusted to and are characteristic of their environmental and geographical context. This they have in common with theories of equilibrium attractors (see below). Davis did recognize different climax forms in environments not dominated by fluvial denudation, but the cycle of erosion predicts the same outcome in all fluvial-dominated settings.

None of these theories are widely accepted now, at least in their original forms, as general models of landscape evolution. But neither are they completely false. Progression toward climax communities and zonal soils does happen, and there exist regionally characteristic soil types and plant communities. These concepts are best regarded, however, as particular cases of more general patterns of community development and soil evolution.

If you had an uplifted landscape eroded primarily by flowing water, uninterrupted by further uplift or subsidence and not subject to any major climate or other environmental changes, a cycle of erosion-type denudation sequence is indeed what you would get. These conditions have rarely, perhaps never, been fully met over long enough periods of Earth history for Davisian peneplains to form. Erosion and planation surfaces, some created mainly by fluvial denudation, do exist, but none has been convincingly shown to be the result of a Davisian cycle. In the 1950s, Lester King (1907–1989), working in South Africa, developed a quite different cyclical model based more on parallel slope retreat than fluvial downwasting. King's model is also a single-direction, single-outcome construct, ultimately producing what King called a *pediplain*.

EQUILIBRIUM

Equilibrium is one of those words that I wish we could just banish (or least require that it be defined or explained whenever it is used)—not because I object to any specific equilibrium concept but because the word has been so loosely used, variably defined, or undefined, that without context it can be quite misleading. In case you are wondering, sustainability, resilience, and tipping point are also on my word banishment wish list for the same reasons (despite the fact that I have used the latter two in article titles).

Equilibrium has been used to refer to everything from a general, vague notion that things have settled down a bit, to precise mathematical definitions, and everything in between. The word is used here—because, since I am not in charge of words, it is there in the scientific literature and public discourse to stay—to encompass various ideas that the attractor states of landscape evolution represent characteristic forms or states that are adjusted to their environmental context, dynamically stable, and self-maintaining to some extent.

To some degree, equilibrium ideas are linked to human psychology and history. Humans desire some sort of a balance or harmony of nature: a sense of order. Though we may sometimes see equilibria because that's what we want to see, equilibria in ESS are not just a figment of our imaginations. Equilibrium happens, and sometimes equilibrium assumptions about attractors dovetail nicely with empirical observations and theory alike. Equilibrium concepts can also be useful as modeling assumptions or benchmarks, independently of whether they represent truths about real landscapes,.

Gregory Cooper (Washington and Lee University) in 2001 boiled down arguments in favor of equilibrium in the context of the role of density dependence in population ecology to their essence. These can readily be expressed as parallel arguments for equilibrium in ESS in general.[1]

Equilibrium argument 1 starts with the stipulation that relative consistency of forms and patterns exists and that this implies regulation. That this order occurs despite variable environmental conditions means that it is the product of self-formed environmental systems and that the forces that achieve it must be intrinsic to the system. When these forces are undisturbed, they lead to equilibrium, the argument continues. Regulatory control to achieve equilibrium implies sensitivity to changes in form, pattern, and morphology. And finally, steady-state mass flux (i.e., inputs and outputs are at least roughly equal) is the most likely mechanism for achieving these controls.

Equilibrium argument 2 (again, drawing parallels with Cooper) also starts with the axiomatic proposition that order exists. Secondly, random, contingent, uncontrolled entities are destined to be transient and to disappear. Therefore consistent, repeated patterns must be regulated, in the sense of being a mathematical or statistical attractor—that is, something must draw or push them toward these attractor states. Next, as repeated, consistent patterns *just are* equilibrium, most forms and patterns are equilibrium. Because equilibrium and regulation imply steady state, most entities are influenced by steady state. As internal adjustments are the most likely instruments for steady state, then these are important determinants of ESS behavior.

To summarize, argument 1 is as follows:

1. Order exists.
2. Order implies regulation.
3. Similar regulation in different environments requires intrinsic mechanisms.
4. If undisturbed, intrinsic mechanisms lead to equilibrium (i.e., order, regularity).
5. Most likely, mechanisms involve or facilitate steady-state mass flux.

And argument 2 starts and ends the same way:

1. Order exists.
2. Non-orderly entities and phenomena do not persist
3. Order (consistent, repeated, patterns) must be regulated.
4. Order $\approx$ equilibrium.
5. Equilibrium implies steady state.
6. Internal adjustments are the most likely tools to achieve steady state.

Again, these are my rephrasing of Cooper's distillation of pro-equilibrium arguments—neither of us is necessarily supporting them. And note that both indicate that something is regulating or guiding the system toward equilibrium, without specifying what is doing so (mysterious ways, indeed).

Explicit, formal equilibrium ideas in geology and geomorphology were pioneered by US geomorphologist Grove Karl Gilbert (1843–1918) in the late 1800s. American geologist John T. Hack (1913–1991) expanded upon these in 1960 as the concept of "dynamic equilibrium." What Hack proposed is better termed a steady-state equilibrium; the quotation marks are there because dynamic equilibrium is a misnomer from a systems theory perspective. What Hack meant by *dynamic* was that equilibrium does not imply a static landscape where nothing much happens, rather an active, changing landscape, but where changes occur around a (steady-state) equilibrium condition adjusted to its environmental context.[2]

In this formulation, steady state is an attractor, as evolution progresses toward a state of topographic adjustment to the climate and geologic framework with inputs roughly balancing outputs (sediment supplied to a river system approximately equal to export out of the system, for example, or soil formation approximately equal to erosion). Though steady-state equilibrium in geomorphology has been roundly debated, critiqued, and defended, the assumption of a tendency for landscape evolution to move toward or oscillate around this attractor shows up regularly in the literature, right up to the present.

Steady states are also often considered attractors in pedology beyond the general adjustment-to-environment equilibrium assumption of climax or zonal soils. A notion of steady-state regolith thickness has been around since the late nineteenth century. Weathering at the bedrock weathering front declines as overlying regolith grows thicker, the reasoning goes, as the weathering front is increasingly isolated from meteorological inputs and biological effects. Erosion at the surface reduces this effect, stimulating weathering and thus maintaining a steady-state soil thickness. This is not a good representation of the way actual soil and weathering profiles are formed and evolve, but it is a perfectly reasonable reference condition and therefore a useful modeling assumption in many instances.[3]

A steady state can often occur with respect to soil organic matter. Here annual litter inputs are approximately balanced, on an average annual basis, by decomposition. Thus, a relatively constant organic matter content is maintained. There are many exceptions, however, such as peats that continually accumulate organic matter, fire-prone and agricultural soils that are frequently disturbed, and dryland soils with little or no soil organic matter.

In ecology, equilibrium has similar connotations of adjustment to environmental conditions, usually accomplished via ecological filtering. Some organisms are favored, while others are inhibited, by the resources and stresses of the environment. With sufficient time, the reasoning goes, organisms and the abiotic environment should be adjusted to each other. A one-way journey toward a single stable equilibrium state is consistent with succession-to-climax concepts. As in geomorphology, such ecological equilibrium concepts have been critiqued for several decades. More so than in Earth sciences, however, ecological equilibria are now generally considered reference conditions and possible outcomes rather than normative (at least by ecological scientists).[4]

RELAXATION TIME EQUILIBRIUM

When a landscape responds to a change or disturbance, it doesn't take forever. Sooner or later, the response is completed (although often well before then slows to a crawl). Imagine your skin's response to a wasp sting. It hurts a lot at first; the pain gradually subsides, and eventually goes away entirely. You can think of this as your skin restoring a stable equilibrium no-pain state, but just as reasonably you can think of it as the effects of the sting being finite.

When environmental changes and disturbances happen, reactions and adjustments are finite and often decelerate. When these have slowed to a negligible rate or ceased, the adjusted condition is sometimes perceived as a new or restored equilibrium. This signifies, at least implicitly, a perception of equilibria as adjustment or adaptation to the changed environmental conditions and/or or as the completion of a response trajectory. This is relaxation time equilibrium (RTE), a much weaker variety than steady states or characteristic forms. Examples of RTE include a slowdown and leveling off of chemical weathering rates as weatherable minerals are depleted, achieving a stable vegetation composition after a fire, or infilling and reduction of channel size after a decrease in flow inputs.

Attractors framed in RTE terms are imprecise (sooner or later, the response slows to inconsequential rates or stops) or associated with some critical limit such as filling a sedimentary basin, reaching ecosystem carrying capacity, or vegetation canopy closure. In fact, RTE is not really an attractor or goal function in the sense of the other frameworks above. RTE just reflects the fact that sooner or later responses to a change or a disturbance run their course. When relaxation time equilibria are interpreted as, for example, steady states, characteristic forms, or fulfilment of a goal function, it brings these situations into play as attractors.

SELF-ORGANIZED CRITICALITY

Considered to be at the "edge of chaos," self-organized critical (SOC) systems feature local instabilities that function to generate broader-scale order. In SOC systems, local processes interact to make prediction of specific failure events impossible, although their probability distribution (with respect to both magnitude and spatial and temporal frequency) can be obtained. Classic examples and models of SOC in ESS include earthquake distributions, the size and frequency of forest fires, slope failures, and drainage networks.

Mark Fonstad and Andrew Marcus of the University of Oregon in 2003 presented one of the best examples of SOC in nature (vs. modeled or laboratory systems) with respect to riverbank failures. Note that SOC is only one of many concepts of self-organization applied to ESS.[5]

In landscape systems, so-called 1/f noise is frequently present. This means that events of all magnitudes occur and that a log-linear relationship exists between event magnitude and frequency, thereby making some processes or time series apparently scale free (that is, the pattern is more or less the same no matter what the resolution at which you observe or measure it). 1/f noise is closely related to log-linear rank-size distributions and Hurst effects (persistence and "memory") in time series such as floods and stream flows. A wide range of phenomena, including city sizes, urban traffic, earthquakes, tectonic events, precipitation, stream flows, and forest fire sizes have been characterized by 1/f noise.

SOC theory can explain 1/f noise and was originally developed in this context by Danish physicist Per Bak (1948–2002) and colleagues in the late 1980s. Later, in 1996, Bak proposed SOC as a general law of nature. Most of the evidence supporting SOC is in the form of 1/f spectra and log-linear magnitude-frequency or rank-size distributions. These can also be generated, however, by processes other than a tendency for systems to evolve to a critical state.[6]

Though SOC theory originated in physics rather than Earth sciences, the seminal and canonical example involved numerical models and laboratory experiments on avalanches in simple sandpiles, facilitating applications to geomorphology. However, real sand landforms, and even realistic laboratory sandpiles, have given mixed results.

1/f noise can also be accounted for by cascades of large-scale fluxes to successively smaller scales. Storage effects complicating input-output linkages in hydrological and geomorphic systems are also a highly plausible and realistic explanation for Hurst effects and 1/f noise in landscapes. Over long geological timescales, long periods of relative stasis interspersed with shorter bursts of intensive activity occur. If you model this kind of tempo using a mathematical tool called a Cantor dust, this can produce the fractal topography exhibiting 1/f noise often found in terrain models and the temporal sequences of uplift and denudation events in the geological record. Log linearity and 1/f noise in landscapes are widely accepted as a signature of self-organization. However, it is a necessary but by no means sufficient condition for SOC. Several different processes may leave such a signature—an example of equifinality.

Note that the relationship between criticality and thresholds may only work on one side of the threshold. Take a hillslope near its angle of repose (the steepest angle that can be maintained in a given material). A slope becoming steeper may indeed reach a critical state whereby a series of failures of various sizes occurs, restoring the threshold angle. But a slope well below this threshold will not steepen toward the critical state.

An instructive example is the work of Jacky Croke (Queensland University of Technology) and colleagues, who examined a large data set of riverbank mass failures for evidence of SOC, publishing their results in 2015. Their results were mixed in terms of a clear yes or no for SOC. For one thing, their Lockyer Creek (Queensland, Australia) study site is typical of landscapes and ESS (as opposed to laboratory and mathematical models), with multiple modes of adjustment, a variety of feedback mechanisms, and local spatial variability of the factors influencing thresholds. Determining a critical state ain't easy in those conditions. Jacky and company also found that fitting their data into an SOC framework was strongly dependent on interpretations and inferential arguments—again, typical of real-world situations. This is not to say that SOC was not helpful, however. By investigating its possible existence, they discovered useful results and insights that they reckoned would not have suggested themselves otherwise.[7]

In ecology, problems of equifinality and interpretation are, unsurprisingly, also common with respect to evidence of SOC in ecological phenomena. An edited volume put together by Jianguo Wu (Arizona State University) and others in 2006 provided a broad overview of scaling in ecology, where power law and log-linear scaling are not uncommon—again for a variety of reasons, SOC being only one of them. As these are in many instances the main or only criterion used to identify SOC, equifinality is an issue in ecology as well as geomorphology.[8]

ALTERNATIVE STABLE STATES

Go out to the nearest playground with a couple of kids and invite them to play on the seesaw. You'll notice that with or without kids on it, it has two stable situations—with the right side on the ground and the left elevated or vice versa. Anything in between is transitory, and the seesaw at rest will be in one or the other stable states, and a seesaw in use will oscillate between them.

The seesaw is a pretty good metaphor for alternative stable-state concepts (I would use the ASS acronym, but that would be in poor taste). In alternative stable-state systems, two roughly oppositional stable states exist, and in-between states are unstable.

Typically, a critical threshold or tipping point separates movement toward one or the other of the alternative stable states—like the literal tipping point of a seesaw.

A canonical example is the trophic and turbidity status of shallow lakes. Some of these are prone to tip between alternative states of clear water, aquatic macrophyte vegetation dominated, and relatively low nutrient status vs. turbid water, phytoplankton dominated, and nutrient overload (eutrophic). Marten Scheffer of Wageningen University and colleagues were able to show, beginning in the early 1990s, via empirical studies and models, how changes in key driving factors push the lakes from one state to the other, with transitional conditions being unstable and short-lived. Since then, other alternative states in lakes have been identified, driven by changes in nutrient inputs, lake morphology, and climate.[9]

Figure 5.1. Seesaw metaphor for alternative stable states. Middle photo is a desert landscape in South Australia showing the alternative stable states of erosion-prone, nutrient-poor bare ground and vegetated non-eroding fertility islands. Bottom photo is a longleaf pine forest in South Carolina showing alternate states of recently burned (left of fire line) and unburned vegetation.

Another well-known example is the relationship between erosion and vegetation cover in many semi-arid environments. These tend to diverge, at the patch or landscape scale, into one of two contrasting states. One is characterized by vegetation cover, a significant stock of soil nutrients, and little or no erosion. The other is a bare-soil, eroded state. The details vary in different locations and situations (of course), but if you think of vegetated islands with thicker, richer soils (at least by dryland standards) surrounded by barren ground you won't be wrong. The specific processes and dynamics were worked out mainly in the 1990s. Earlier, and often unrecognized, the redoubtable John Thornes in 1985 showed that under realistic assumptions and common circumstances, the interplay between vegetation cover is an alternative stable-state system. Contrasting states of fully eroded with no vegetation or complete vegetation cover and no erosion are stable. In-between states—various combinations of eroded bare soil and vegetation cover—were found to be unstable, with small disturbances tipping the system toward one extreme or the other. Thornes was a pioneer in linking observed ecological, geomorphological, and hydrological phenomena to the dynamics of nonlinear dynamical systems.[10]

Well known to, but widely misunderstood by, geomorphologists, hydrologists, and engineers is something called the Lane Diagram, first presented in 1955 in a publication by E. W. Lane (1891–1963). The Lane Diagram (Figure 5.2) describes an alternative stable-state system, though it is widely (mis)interpreted as a steady-state attractor.[11]

The diagram shows that stream degradation (net erosion and incision) and aggradation (net deposition) respond to changes in the relationship between sediment supply (amount of sediment, Q_s, and typical sediment size, D_{50}) and sediment transport capacity (a function of discharge or flow, Q_w, and slope, S). When sediment supply outweighs transport capacity, the balance tips toward aggradation. If transport capacity exceeds sediment supply, it tips to degradation. The diagram is a very helpful metaphor in understanding the sediment supply vs. transport capacity relationship

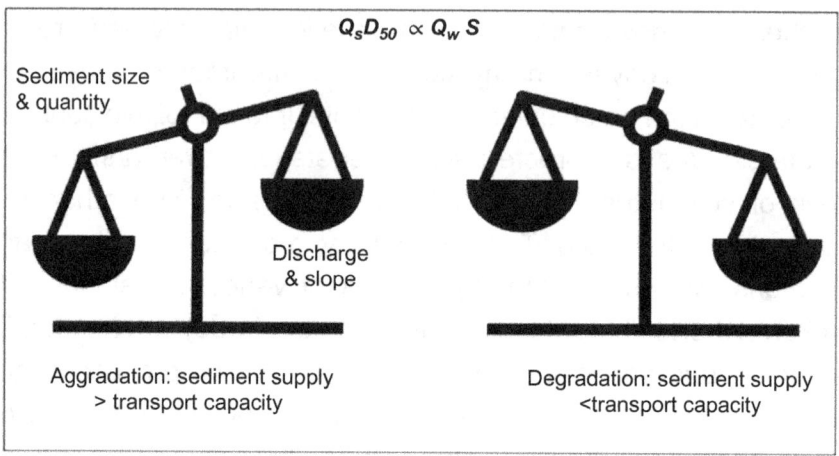

$$Q_s D_{50} \propto Q_w S$$

Sediment size & quantity

Discharge & slope

Aggradation: sediment supply > transport capacity

Degradation: sediment supply <transport capacity

Figure 5.2. My version of the Lane Diagram for fluvial systems (see text for explanation).

and its effects on channel aggradation or degradation—one could imagine instead of a balance, sediment supply and transport capacity being loaded onto one side or the other of a seesaw.

Somehow, though, through no fault of Lane's, the balance diagram began to be interpreted as representing a tendency of rivers to keep the balance balanced—as though a seesawing river would for some reason seek to maintain a delicate balance at the tipping point. As Q_s, Q_w, D_{50}, or S changes and tips the balance (the story goes), the stream adjusts one or more of the other factors to return to the middle, balanced condition—a steady state where sediment supply equals sediment transport capacity.

What the Lane relationship really tells us is that if the system *is* precisely balanced (in steady state), a change in any factor will tip the scales one way or another. It also tells us that if the scale is way overbalanced on the aggradation or degradation side, it may take a lot of change to level it out or tip it back the other way. The Lane Diagram is therefore better seen as an alternative stable-state system, whereby the balance point is unstable, and the system tends to tip toward the (stable) aggradation or degradation states. Indeed, many rivers are undergoing net aggradation or degradation, while steady state is relatively rare and transitory.

The seesaw analogy applies only to binary alternative states and when the alternative stable states are defined in terms of specific criteria—for instance, oligotrophic or eutrophic lakes or aggrading vs. degrading fluvial channels. There are, of course, many other important characteristics of streams and lakes. Alternative stable states may also be scale contingent. Unlike a seesaw, where states change in a matter of seconds, some landscape systems may take decades, centuries, or longer to transition between states. There-fore, on a time frame of (say) months or years, the alternative states may be undetectable or irrelevant.

An old joke holds that there are two kinds of people in the world: those who divide the world into two kinds of people and those who don't. Some of the binary alternatives are certainly real; they accurately reflect the way real landscapes work. Others may arise from the human tendency to see things in terms of binaries and dichotomies—I can't give you a convincing example, I'm just allowing this as a possibility. In many other cases, the two alter-native stable states separated by an unstable threshold are likely embedded into more complex patterns of state transitions.

The ASS (sorry, I had to do it just one more time) framework is sometimes a reasonable way to analyze or represent whole-system behavior and sometimes an essentially reductionist way to simplify landscape systems. Either way is okay—and looky there, I just spun another binary.

At any rate, where more than two possible stable or at least persistent states exist, seesaws or balances don't cut it, metaphor-ically speaking.

MULTIPLE PATHS, MULTIPLE OUTCOMES

So how about those situations where >2 outcomes are possible from similar starting points? This is common in landscapes, leading to complex geographical patterns and extensive spatial variability over small areas (Figure 5.3).

Figure 5.3. Examples of complex geographical patterns, with extensive variation over short distances and small areas. Left: ecological systems in a 100 square km area of central Texas (center coordinates 29.7° W, 90.50° W; approximately 100 km west of Houston). The area contains thirty-three separate ecological systems (derived from the Texas Ecosystem Analytical Mapper, https://tpwd.texas.gov/gis/team/). Right: field soil mapping in the Ouachita Mountains, Arkansas. Detailed soil mapping of 0.13 ha plots showed four to ten different soil series per plot and different series in 60 percent of sample pairs ~ 1 m apart.

Sticking with the game metaphors, let's make a domino chain (Figure 5.4).

The domino chain mimics a complex, interlinked, many-component landscape. Transgressing a local threshold—i.e., tipping one domino—causes some or all the others to fall as well. But exactly how many dominoes fall, and in what direction, is sensitive to the location of the domino in the line and which way it is tipped. Unless you tip one of the blocks on either end of the chain to the outside, each tip directly knocks over one other. The end dominos are all or nothing: if they tip one way, nothing else happens; the other way, and every other domino falls. For all the others, the number of dominos that fall also depends on which way an individual tips. Any domino is highly sensitive to any left tips of any dominoes to its right and insensitive to any left tips to its left, and vice versa. In the *Landscape Evolution* book, I presented a probabilistic mathematical model of this (because that's what people like me do), but it is simple enough to visualize. Even when you rule out breezes, earthquakes, and grandchildren (mine never saw a domino they did not want to knock over) and confine the possibilities to what happens when you tip a single domino right or left, it is simple to see that what happens is predictable if you know exactly which block is tipped and in which direction— but you must indeed know *exactly*. You cannot make a prediction of the outcome of one tipped domino that would generally apply.

Figure 5.4. Dominoes.

Likewise, a pattern of fallen and standing dominos allows you to figure out the position and tip direction of the initial falling piece—that is, assuming it was by a single tipped domino. For a disturbance affecting multiple dominos, all bets are off. The domino metaphor also illustrates the fact that if a disturbance is big enough (such as shaking the table, a blast of wind, or a mischievous five year old) it resets the entire system.

In summary, the domino chain metaphor illuminates several key properties of many landscapes:

- Sensitivity to small disturbances (one tipped domino)—but not in every case
- Disproportionate responses (poking one block can cause many more to fall)—but not in every case

- Alternative stable states with an unstable threshold between them (an unstable standing domino separating stable states of lying on its left or right side) embedded within a broader system of many potential outcomes
- Subject to unbreakable scientific laws—the Newtonian physics of tipping—but not predictable solely on the basis thereof
- Deterministically predictable in some situations (single initial tipped domino), but only with exact information (which domino and which direction)
- Potentially subject to complete resetting by changes or disturbances of sufficient magnitude
- Changes that are irreversible without outside help

Now let's look at some conceptual frameworks linked to multi-path landscape evolution.

UN-EQUILIBRIUM

Geomorphologist Bill Renwick of Miami University in a 1992 article articulated a key distinction between (true) nonequilibrium and disequilibrium. *Disequilibrium* systems are progressing toward a stable equilibrium state but haven't had enough time to get there. This could be because the landscape is relatively young or due to disturbances. Disequilibrium implies the existence of one or more stable equilibria and is not inconsistent with single-equilibrium or alternative stable-state concepts. However, some ESS are so frequently disturbed that they can be pretty much perpetually in disequilibrium—in effect, the disequilibrium state is the attractor. Some examples include fire-prone and fire-dependent ecosystems (Figure 5.5), barchan dunes (crescent-shaped dunes common in some deserts), gravel riverbeds, and frequently overwashed coastal environments.

Nonequilibrium ESS are not characterized by a path toward any stable equilibria, with or without disturbance. Nonequilibrium

Figure 5.5. Prescribed fire is often used to mimic natural fire regimes in fire-dependent ecosystems like this one at Modoc National Wildlife Refuge, California. U.S. Fish and Wildlife Service.

systems may be dynamically unstable and deterministically chaotic and thereby sensitive to small perturbations (metaphorical domino fall) or minor variations in initial conditions. In other cases, nonequilibrium systems may just lack the tools for achieving equilibrium. For instance, achieving steady state in alluvial channels is based on iterative adjustments of mechanisms for dissipation of excess energy. But not all streams have excess energy to be dissipated and may thus be nonequilibrium. Adjustment of biological distributions to climate change may be nonequilibrium because the organisms lack effective dispersal mechanisms or encounter barriers to movement.

Nonequilibrium in dynamical systems terms involves *unstable attractors*. Systems are often analyzed in dynamical systems theory using a multidimensional phase space defined by two or more key variables of the system. An attractor is a portion of the phase space that all evolutionary trajectories eventually move to (in mathematical modeling produced by numerical simulations). An attractor may be a single point or small area of the phase space, and some are periodic, featuring oscillation between points within the attractor. Other attractors are more complex and are called strange or chaotic attractors. A butterfly-shaped attractor produced by a simplified three-variable version of the equations of motion for the atmosphere is the best known. It is called the Lorenz attractor, introduced in a 1963 paper by meteorologist Edward Lorenz (1917–2008) and often cited as the birth of chaos theory. Lorenz showed that the system diverges over time from even infinitesimally different starting points. But the divergence is bounded and non-random—the system evolution is drawn to the complex attractor shown in Figure 5.6. All model trajectories move to, and thereafter stay in, this portion of phase space, though no trajectory ever repeats exactly.

Figure 5.6. The Lorenz attractor (see text for explanation). Source: Wikimedia commons; public domain.

Deterministic chaos is, in most instances, equivalent to dynamical instability. Chaos and instability have been demonstrated for many ESS and associated processes. Many of these, however, are based on simulation models or, at best, number crunching of time series data of, for example, stream flows, precipitation, seismic signals, or population fluctuations. Relatively few are based on or supported by field evidence, and fewer still are directly linked to landscape evolution rather than shorter-term phenomena. For example, Bellie Sivakumar of the Indian Institute of Technology, the foremost analyst of complex nonlinear dynamics in hydrological phenomena, has demonstrated deterministic chaos in multiple cases for hydrological phenomena such as stream flow and rainfall (as have others). However, that does not necessarily translate directly to a chaotic attractor in the evolution of, say, preferential flow in soils, fluvial networks, or groundwater flow systems.

Studies showing dynamical instability and chaos directly in landscape evolution phenomena certainly exist—I've published some myself—but these do not and cannot yield the mathematically precise and visually fascinating patterns of the Lorenz attractor or anything like it. And unlike mathematical models, actual landscape evolution deals with attractors, strange or otherwise, that are not truly final states.

For an example, let's return to the mutually adjusting relationships between channels and the flows they convey (hydraulic geometry, discussed in Chapter 1). At-a-station hydraulic geometry addresses how different flows are accommodated at a fixed location along a channel by variations in velocity, width, depth, slope, and frictional resistance. The interactions of these factors in standard physics-based flow equations show that these relationships are dynamically unstable. The instability is expressed not as an attractor in phase space but as multiple modes of adjustment—different combinations of increases and decreases of velocity, width, depth, slope, and frictional resistance in response to incoming flows.

The multiple modes of adjustment include instances of opposite-from-expected behavior. The expected behavior, according to basic physics (and intuition), is that when discharge decreases, velocity, width, depth, and slope would also decrease, and roughness or resistance should increase—and vice versa for an increase in discharge. But an increase in imposed flow, for example, could be accommodated by increases in depth and decreases in resistance (for instance) that could convey the flow even with an opposite-from-expected *decrease* in velocity. Opposite-from-expected behavior has been documented by examining changes in the hydraulic variables at a cross-section as different imposed discharges pass through. Multiple modes of adjustment and occasional opposite-from-expected changes also apply to downstream (as opposed to at-a-station) hydraulic geometry, which deals with the along-channel changes in channel and flow characteristics, resulting in extensive local variability superimposed on broader general upstream to downstream patterns.

These complicated changes in hydraulic geometry are not attributable to randomness or stochasticity. They arise from the fundamental interactions among the hydraulic variables and are therefore *deterministic* complexity. Possible responses of channels to changes in flow or of flow to channel changes are constrained by local place factors (such as the resistance of the bed and banks or geological constraints on slope) and by general laws. The laws decree, for instance, that if velocity does increase or fail to decline in response to lower discharge, one or more other factors must adjust to accommodate the change. Despite the instability and seemingly pseudo-random variability of hydraulic geometry, the fluctuations occur within well-defined, albeit relatively broad, boundaries, and broad general patterns and trends can be predicted. For example, in a typical river where discharge increases upstream to downstream, channel width (for instance) may vary extensively from one cross-section or reach to the next, but on average the channel will get wider as you get farther downstream.

RESPONSE OPPORTUNITIES AND CONSTRAINTS

In their journey toward whatever is next, landscapes respond and sometimes adapt to changes. Just as there are several different ways for desert plants to adapt to dryness, and for mollusks to adapt their shells to the demands of their habitat, there are different ways for landscapes to adjust. However, just as a snail does not think about how to modify its shell morphology to avoid getting swept away by currents or to minimize the risk of getting eaten by a predator, landscapes do not strive toward any particular response or adaptation. Responses happen, and the ones that increase stability and efficiency tend to persist—just as genetic variations that make for better mollusk shells are favored by natural selection.

So how many options or possibilities for response are there?

PLASTICITY AND DEGREES OF FREEDOM

The form of plasticity most familiar in ecology and biogeography is phenotypic plasticity, the ability of an organism's phenotypes to adapt to the environment. I first began thinking about this in a landscape concept after encountering the work of Liang Liang (who occupied the office next door at the University of Kentucky), who studies phenotypic plasticity in the context of plant responses to variations or changes in climate. Plants have differing abilities to adjust their phenology (for instance, the timing of flowering, budding, leaf-out, senescence) to climate. Plasticity is generally considered in human geography and economics as the range of options available to individuals, households, states, or firms to respond to external changes and adapt to their broader environment. Plasticity in landscapes is closely related to *degrees of freedom*.[12]

If you had a statistics course along the way, you probably think of degrees of freedom by its statistical meaning, the number of values that can vary. In dynamical systems theory, the number of degrees of freedom is how many independent ways in which a dynamic system can move without violating any imposed constraints. I am using the term here in a similar way—the number of different mechanisms by which a landscape could adjust to changing conditions. In the hydraulic geometry example above, for instance, there are five degrees of freedom for responding to changes in flow: width, depth, velocity, slope, and frictional resistance.

Think about how a landscape can handle rainfall inputs. All the water must go somewhere and do something, and it can be evaporated, used by plants and transpired, stored in the soil, stored at the surface (depression storage such as puddles), percolated downward to groundwater, or it can flow away as surface or subsurface flow. That's seven degrees of freedom. But now consider wet conditions—the degrees of freedom and the amount of water that can be handled by each may be limited by soil thickness and moisture storage capacity, hydraulic conductivity (rate at which water can move through the soil), vegetation cover, ground slope, and antecedent conditions (i.e., how wet was it to start with). If the soil is already saturated, for example, infiltration is not an option, and the degrees of freedom linked to it (soil moisture and groundwater storage, subsurface flow) are not available. Plants and soil moisture have first dibs on rain falling on dry soil; therefore, percolation to groundwater may be not an option. Surface runoff cannot occur unless certain conditions are met (precipitation intensity faster than infiltration or surface saturation). Obviously, the response options can be quite variable from one rainstorm to the next, even within a single landscape. Over landscape evolution timescales, responses to chronic excess moisture inputs could include increasing plant cover (plants using more water), thickening of soil profiles due to pedologic processes associated with water inputs, and the creation or enlargement of surface/subsurface channels or flow paths. A drier regime would potentially be reflected in the amount and type of vegetation, changes in soil properties (for instance, reduced translocation and increasing retention of carbonates and salts), and abandonment or deterioration of channels.

The lessons are that degrees of freedom are often numerous, available responses are variable even within the same landscape, and without quite specific event- and site-specific information, landscape responses are extremely difficult to predict. However, the degrees of freedom are finite, the response space is constrained, and the range of possible environmental responses and their relative probabilities *can* be determined.

Now, more about. . . .

CONSTRAINTS

Constraints limit the way a landscape can respond to changes. These are often linked to laws such as mass and energy conservation that cannot be violated or place factors that limit the resources on hand to enable or fuel responses. These limiting factors often relate to fundamental limits such as geographical boundaries, maximum photosynthetic efficiency for plants, carrying capacity for ecosystems, the base level of erosion, and so on.

The boundaries or limits within which landscape evolution occurs can be thought of as an *evolution space*. This is a multidimensional space defined not only by geographical or geometrical space but also by the available energy and mass for landscape evolution, as well as necessary elements such as water or nutrients. An ecological niche is often thought of, not incorrectly, as the role an organism plays in an ecosystem. But as originally conceived by George Evelyn Hutchinson (1903–1991) back in 1957, the niche is a hypervolume (that is, a multidimensional space with >3 dimensions) defined by geographical space and the range of necessary resources such as sunlight, water, and nutrients. While Hutchinson and most ecologists think of niches at the individual level, Charles Smith (in the 1980s) and Edmundas Lekevicius and Andrei Lapenis (in the 2000s) envisioned this multidimensional space at the scale of ecological systems rather than individuals or taxa. I adapted their ideas to the notion of a *resource space* based on the availability of energy and matter resources and geographical space, all of which can be modified by organisms within the system or by external changes. An analogous geomorphological landscape evolution space can be defined based on available energy and mass available for geomorphic processes, again modifiable by external factors as well as the development of the landscape itself. The true landscape evolution space would include both the ecological resource space and the geomorphological evolution space (at least).

The key idea, boiled down, is that while many possibilities exist in landscape evolution, they must occur within a defined, constrained evolution space. *Many* things can happen, but not *anything* can happen.

GOAL FUNCTIONS AND EMERGENCE

Increasing fitness of landscapes implies that ESS components become increasingly adapted to their environmental context and to each other. The term *fitness* is rarely used to describe evolution in geomorphology, pedology, or hydrology, however, except with respect to how these factors are linked to evolutionary biological fitness. Most folks are familiar with the phrase "survival of the fittest" as a bumper-sticker summary of Darwinian evolution (though the phrase is due to Herbert Spencer; Darwin did not use it in his writings). Fitness is defined in evolutionary biology as organism traits that tend to increase reproductive success of individuals. Use of the term in ecology pretty much parallels that in evolutionary biology, but ecological fitness is sometimes perceived more broadly to refer to how well organisms or taxa are adapted to their environment in terms of stress tolerance, resource acquisition, habitat availability, and other factors that influence reproductive success. The "fitness" of geomorphic, soil, hydrological, and ecosystems (with or without use of that specific word) is typically expressed in terms of stability or optimality.

EXTREMAL HYPOTHESES AND OPTIMALITY

A nitrogen atom doesn't know and doesn't care whether it exists as nitrogen gas, nitrate, nitrite, ammonia, or in gas, dissolved, or solid form. A grain of sand doesn't know or care whether it stays put or gets eroded, transported, or deposited by wind or water.

While sentient organisms exist within landscapes, non-sentient elements do, too, and landscapes themselves are not sentient—they just don't care.

Development of ESS is characterized, or even governed, by a tendency to maximize or minimize some aspect of mass/energy flux or internal organization—that's the basic idea of extremal hypotheses, and there are many. These maxima or minima are held to optimize the functions of ESS, and for brevity are referred to here as optimal principles. To some degree, all the optimality principles summarized below succeed in describing or explaining some aspects of landscape evolution, with some empirical and case study support. But it has yet to be explained why an ESS should seek to maximize, minimize, or optimize anything.

Extremal hypotheses use different terms and methods and emphasize different parts of landscape systems, but optimal principles have been proposed in ecology, hydrology, geomorphology, climatology, and fluid dynamics. Many of these, if not equivalent, are at least mutually consistent. Optimal principles based on four different thermodynamic criteria proposed to govern ecosystem development, for instance, are all mathematically related, and their measures are strongly correlated. Most optimal principles proposed for ecosystems are mutually consistent with respect to their fundamental implications, which imply that work is performed as efficiently as possible, consistent with least work and maximum entropy production principles.[13]

Thermodynamics-based optimality principles are based on throughputs, storage, and transformations of energy and mass. Without getting too far into the weeds on this, these mainly boil down to variations on the principle of least work and its corollary, maximum entropy production. These hold that a system does whatever it does (move fluids, transfer heat, grow biomass, transport sediment, etc.) as efficiently as possible given the available energy. Because of the law of energy/mass conservation, the less energy used to get the job done (least work) the more is dissipated in other ways (maximum entropy). In addition to ecological systems, maximum entropy or least work principles have been shown to apply to purely or mainly abiotic phenomena such as heat flux in the atmosphere and oceans, global climate, fluid convection, turbulent dissipation, and hydraulic geometry (the mutual adjustments between channels and the flows they convey). In many cases, these have been explicitly shown to be consistent with least work / maximum entropy production.

Confusion sometimes arises because some of these optimal or extremal principles are phrased in terms both maximizing or minimizing energy, entropy, or work. The key is in *what* is being purportedly maximized or minimized. The minimum sediment transport hypothesis for streams, for instance, does not propose that sediment transport itself is minimized, but rather that the energy used per unit of sediment transport is minimized.

Likewise, maximum entropy principles for ecosystems do not suppose that increasing the amount of energy that does not perform work is somehow advantageous, but rather that the work of, for example, net primary productivity or water use is performed using the minimum proportion of available solar energy, thus increasing the proportion dissipated as entropy. For a given input of energy, maximum efficiency in accomplishing work, coupled with conservation laws, dictates maximization of energy dissipation via entropy (maximum entropy production, MEP)—thus, there exists general consistency of optimality principles based on energy, power, and entropy.[14]

In physics, the least work criterion is often called the least action principle (LAP). It holds that in a conservative dynamical system the motion between any two points is such that the action has a minimum value with respect to all paths between the points that correspond to the same energy. The LAP implies that nature finds the most efficient path, such as accomplishing work in landscapes using as little of the available energy as possible. In the table below, some optimal principles are directly linked to the LAP by proposing maximum efficiency in energy use and/or mass fluxes. Others are based on a maximum entropy principle or propose maximum energy throughput, which also implies MEP. A third group is based on preferential utilization, preservation, or replication of the most efficient flux gradients. While not directly expressed in thermodynamic terms, these are completely consistent with LAP and MEP.

This is not to say that these principles are redundant—among other things, they have different applications, are derived from or tailored to different phenomena, and have varying criteria and metrics. Many are useful and interesting independently of their commonalities. But *why* do these principles work in so any cases? Ecosystems, watersheds, and soils, for instance, may be far from the ideal conservative dynamical systems specified in the LAP in physics. Landscapes cannot plan or desire any specific pathway or outcome.

If we rule out goal functions, there remain at least three potential explanations for this common behavior (I call it *phenomenological equifinality*). First, pathways and outcomes associated with the optimality principles are more probable than other outcomes. They are not preordained or dictated by any general laws, but their consistency with more efficient energy/mass fluxes makes them more likely to occur. Second, when optimal processes and structures happen to emerge, positive feedbacks reinforce them. Third, features and evolutionary pathways associated with optimality are favored by selection processes.

The answer: all three. Optimal/extremal principles generally assume that optimal outcomes are probabilities or tendencies, not inevitable or deterministic. The mechanisms held to create these probabilities are based on feedbacks that may reinforce optimal phenomena that happen to occur or mitigate against suboptimal trends. The probability and feedback explanations can therefore be lumped into *selection* processes.

EMERGENCE

Again, why would a grassland, a soil, a stream, or a mountain slope maximize, minimize, or equalize any quantity or flux? They are indifferent and lack self-awareness. Certainly, an individual organism can benefit from doing so, but communities and ecosystems? And non-living rocks, minerals, water, etc.? Questions of this sort are why many scientists are skeptical of metanarratives based on "balance of nature" ideas whereby ESS are supposed to seek some form of balance or equilibrium. Explanations based on such (apparent or purported) goals are much more plausible if they can be framed in terms of emergent behavior. Emergence is independent of (though not necessarily at odds with) any teleological implications and is simpler than postulating goal functions.

The only thing necessary (and sufficient) for optimality to emerge is that higher probabilities of optimal behavior are associated with positive feedbacks. That is, when the optimizing trend or action does occur, it is reinforced. No principle of optimality is required. Optimal-like behavior in landscapes therefore doesn't need, or even imply, any goal functions. If the maximization or minimization or balancing involved improves the likelihood of survival and replication of the responsible entity, progress toward optimality occurs. Optimality in landscapes is not deterministically inevitable or promoted by any guiding entity; it is an emergent property arising from selection.

Take, for instance, theories and hypotheses of a tendency toward increasing productivity in ecological systems (see Tables 5.1, 5.2). In a multidimensional resource space allocated among organisms, just three postulates produce a developmental pathway toward increasing productivity. First, there exist advantages in survival/reproduction associated with higher rates of resource procurement and use. Second, the environment is biologically saturated or tending toward saturation—that is, the resource space is filled or being filled. Third, the resource space is not shrinking due to external factors. That's it. If these common conditions are

Table 5.1. OPTIMAL PRINCIPLES AND GOAL FUNCTIONS BASED ON, OR IMPLYING, MAXIMUM ENTROPY PRODUCTION (MEP)

Principle	Applies to . . .	Summary	Comments	Date
Maximum power	Ecological systems	Systems organize to maximize energy throughput	Equivalent to maximization of energy throughflow	1922
Maximum generation of available potential energy	Atmosphere	Atmosphere heat flux operates to maximize rate of potential energy production	Equivalent to maximum entropy production	1960
Maximum energy cycling	Biological systems	Systems organize to maximize mass and energy cycling		1968
Minimum entropy exchange	Climate	Ocean-atmosphere heat flux min-imizes entropy exchange with external environment	Equivalent to maximum entropy *export* to external environment	1975
Maximum exergy storage; maximum emergy (embodied energy)	Ecological systems	Systems maximize storage of useful energy (emergy)	Accumulation of mass and exergy; exergy = maximum possible useful work	1979
Maximum energy residence time	Ecological systems	Systems organize to maximize energy residence time	Equivalent to maximum exergy storage and emergy	1981

Table 5.1. Continued

Principle	Applies to . . .	Summary	Comments	Date
Maximum energy dissipation	Biological systems	Systems increase order at the expense of disorder (entropy) in surrounding systems	Broadly equivalent to maximum entropy production	1994
Increasing energy flow	Biosphere evolution	Biosphere has evolved by maximizing energy flow	Self-organizing mechanisms promote maximum energy efficiency	1999
Maximum energy flux	Ecohydrology; plant water use	Natural selection favors maximum energy flux	Maximum equated with optimum	2002
Maximum entropy production (MEP)	Turbulent flows	At steady state, entropy export maximized by turbulent energy dissipation		2003
MEP	Plant physiology	Optimization theories unified by MEP	"Survival of the likeliest"	2010
MEP	Environmental and ecological systems	Nonequilibrium thermodynamic systems organized in steady state such that entropy production is maximized		2010

MEP	Vegetation and carbon assimilation	Vegetation evolves toward maximum productivity, associated with MEP		2012
Maximum power	Drainage basin evolution	Maximization of sediment transport to deplete topographic gradients	Based on tendency of ESS to deplete driving gradients as rapidly as possible	2013
MEP	Ecological succession	Rate of entropy production increases during succession	Ecosystem net energy budget must export entropy	2015

Date refers to the first formal proposal in the scientific literature. Supporting references are given in *Landscape Evolution*, Chapter 6.

Table 5.2. PROPOSED OPTIMALITY PRINCIPLES BASED ON THE PRINCIPLE OF LEAST ACTION AND/OR GRADIENT SELECTION

Principle	Domain	Summary	Comments	Date
Maximum energy efficiency	Biological systems	Evolution selects for most energy efficient combination of organisms	Focuses on mutual aid in evolution rather than competition	1902
Minimum stream power	Fluvial channels	Channels adjust to transport sediment with minimum possible expenditure of work	Phrased in various forms	1967
Maximum flow efficiency	Fluvial channels	Channels adjust to maximize flow efficiency and minimize energy expenditure	Phrased in various forms	1980
Increasing ascendency	Ecosystems	Ecosystem development characterized by increasing ascendency	Ascendency a function of total mass/energy flux and specificity of each flow	1980

Minimize empower/exergy ratio	Ecological systems	Efficiency enhanced by maximizing empower relative to exergy	Empower = rate of emergy acquisition	1997
Least action	Fluvial channels	Channels tend to adjust to transport sediment with the minimum possible work	Consistent with maximum flow efficiency	2000
Gradient selection	Geomorphic systems	Steeper, more efficient flux paths tend to persist and grow	Also: resistance selection—preferential preservation of more resistant features	2010

Date refers to the first formal proposal in the scientific literature. Supporting references are given in *Landscape Evolution*, Chapter 6.

met, the system will evolve along a path toward maximum energy and matter fluxes, storages, transformations, or cycling. No goal functions required.

Let's look at an abiotic example to see whether emergence and selection operate without biota involved: barchan dunes. These are crescent-shaped sand dunes with the horns pointing downwind (Figure 5.7). In sandy arid zones where sand supply is limited, sediment-transporting winds are dominantly from a single direction, and vegetation is absent or minimal, barchans occur.

Figure 5.7. Barchan dunes of Mui Ne, Phan Thiet, on the coast of Vietnam.

Barchans have been studied and explained in many cases based on steady state, whereby losses and gains of sand are roughly balanced and size and shape remain constant even as the dune migrates downwind. That is, by implication if not assertion, the barchan crescent is an optimal shape that maintains a steady-state balance and allows for maximum rates of movement. Alas, empirical studies and theoretical models have shown that size and shape invariance are not necessarily all that common. As winds and sand supply change, the dunes undergo shape transitions rather than maintaining or restoring their morphology (though some degree of shape consistency exists, indicating the classic crescent shape is an attractor).

A pile of sand blowing in the wind does not give a damn what shape it takes or how fast it moves. But cellular automata models of aeolian dunes with only three simple rules can produce common dune forms, including barchans, when the necessary sand, wind, and vegetation conditions are present. We need only stipulate that sand transport by saltation occurs only in the direction of the wind, that deposition of transported sand is more likely where sand already exists, and that an angle of repose exists (which it does) so that when sand elevations increase to where the angle of repose is exceeded, avalanching occurs until the critical angle is obtained. These basic rules, based on standard principles of aeolian geomorphology, do not imply anything at all about dune morphology. Yet, under the stated wind, sand, and vegetation conditions, they indeed produce barchans—without, by the way, implying or stipulating steady state.[15]

MULTIPLE CAUSALITY

I cannot claim that emergence and selection are always sufficient to explain landscapes or the components thereof. Some things that occur or recur around the planet have causes that are so different in different places that emergence and selection are difficult to apply, at least directly.

An example that I have worked on is vertical texture contrast soils, where coarser surface layers overlie finer-textured subsoils. In my yard, for example, loamy sand surface horizons cover sandy clay loam illuvial horizons. These texture contrast soils are ubiquitous around the world. Traditionally, and accurately enough in some cases, these have been explained by vertical translocation by infiltrating and percolating water. Water moving downward through the soil washes smaller, mostly clay-sized particles out of the top layers, and it accumulates lower down (a process called lessivage or argilluviation). Or substances dissolved by percolating water are removed from the upper layers (where the insoluble material is often dominated by larger quartz sand grains) and chemically precipitate in the subsoil.

But, multiple causality. Other processes can account for vertical texture contrasts within a given soil (i.e., multiple causes at work) and between soils in different places. For instance, in an east Texas county, I found evidence supporting and refuting five general types of explanation for vertical texture contrasts. Sometimes multiple causes are necessary to account for them. In some coastal plain sites I studied in North Carolina, vertical translocation occurs, as does faunalturbation and floralturbation (soil mixed by animals and plants, respectively). While all are viable explanations for texture contrast soils in some settings, at these locations none can produce the observed texture contrasts without the other two.

In general, soil layering can occur due to inherited sedimentary layering; deposition of coarser surface material by wind, water, or mass wasting; episodic surface erosion, deposition, and stability; erosional winnowing (preferential removal of finer material from the surface); weathering; vertical or lateral translocation; bioturbation; and various combinations of these.

Even in this case, though, we can propose a rough emergent explanation based on just one overriding principle: smaller or less-resistant material is more likely to be moved downward, downslope, or downwind. Thus, more soluble or finer materials have a greater tendency to be translocated downward or preferentially removed from the surface. In either case, the surface layer is coarsened, and in the case of translocation, the finer stuff accumulates in the subsoil. Bang: vertical texture contrast.

Despite this multiple causality and equifinality, some general principles apply, however broadly. Gravity moves infiltrating water downward from the soil surface to the subsoil. Things that are easier to move because they are smaller or more soluble prefer-entially move with the water. Similarly, water or wind erosion at the surface preferentially removes smaller, lighter particles, coarsen-ing the surface layer. When sediment is deposited at the surface by geomorphic processes or bioturbation, larger material is more likely to remain in place, while smaller (or less-resistant) particles are more likely to be moved downward, downslope, or downwind. Therefore, we could posit an emergent explanation based on the principles that finer and less-resistant material is more likely to move downward in the soil profile, while coarser and more resistant material is more likely to remain in the upper profile. However, any pedologist could cite exceptions: the tendency of rock fragments in some cases to move downward due to gravitational settling and un-dermining by soil fauna, for example, or the persistence of resistant clays in some surficial layers.

CIRCULAR REASONING

Here we take a look at a recurring geometrical phenomenon in landscapes—the circle and its three-dimensional relatives: the sphere, hemisphere, and cylinders. In landscapes, circular shapes are everywhere—animal burrow openings, center-pivot irrigation areas, impact craters (from raindrops to meteors), explosion craters, sinkholes, weathering cavities, tree canopy "footprints" (driplines).

The simplest explanation is that circles and spheres are effi-cient, meaning that they will be selected for. The circle is the 2D shape with the smallest perimeter/area ratio, and the sphere is the most efficient 3D shape for enclosing a given volume. There-fore, an ant or a wombat digging a nest or burrow seeking to get the job done with least effort constructs more or less circular passages and openings. Surface tension acting to pull molecules into the tightest possible grouping forms spheres, and the effects of these spheres (bubbles) tend to be approximately circular or hemispherical (e.g., cavitation pits in rock). Farmers seeking to irrigate the maximum area of cropland with the minimum amount of pipe use the center-pivot system where topography allows it, resulting in circular vegetation and soil moisture patterns. So, are all landscape circles a manifestation of geometric efficiency? Not necessarily.

A point-centered disturbance with no directional bias (that is, no tendency for effects to be significantly greater in any direction away from the point) also produces a circle. Therefore, explosion craters from volcanoes or bombs, and impact craters, are approximately circular. Sinkholes formed by solution centered on an intersection of vertical joints, or a shaft are also a point-centered process, and therefore such sinkholes are roughly circular.

Isotropic dispersion from a point also produces circular patterns. When not affected by other plants or structures, tree branches grow away from the trunk with an equal probability in any direction (same for roots below the ground for trees with lateral roots). Over time, the extent of branches and foliage away from the trunk is approximately equal on all sides, so that the zone of influence on the ground reflected by driplines, litter fall, and soil moisture drawdown is circular. Animals foraging from a central point (nest or burrow) will also produce circular impact areas when resource distribution is isotropic.

Consider also preferential preservation. Sometimes the processes that produce a given form may not necessarily tend toward maximum efficiency, but once formed, those more stable or efficient structures may be preferentially preserved. Thus, for example, weathering cavities on a rock surface with a more spherical (or hemispherical) shape may be more mechanically stable and therefore preferentially preserved compared to other cavity shapes of similar volume.

No single explanation, however general, can account for all or most of the circular features in nature. However, most explanations can be boiled down to efficiency-selection or point-centered processes with no directional bias. Even the latter, however, could be linked to efficiency selection, as (for instance) insect foraging from a nest entrance in a circular area would be the most efficient way of covering area, and ejection of material in a circular area the most efficient way (in terms of mass/area) of disposing of the ejected mass.

CONSILIENCE?

Notwithstanding the uniqueness and idiosyncrasy of landscapes, certain regularities are evident. Soils with vertical texture contrasts (coarser over finer) occur in a variety of parent materials, climates, and topographic settings. Stream channel networks with similar topologies occur in greatly different geological and topographic settings and hydrological regimes. Banded vegetation and soil patterns occur in different environments, glacially eroded topography is similar in mountains of contrasting lithology and tectonic settings, and forest communities tend toward similar compositions across broad regions. Why the repeated occurrence of these features, forms, and patterns? Some possible general explanations are explored below.

Maybe similar phenomena recur due to pure *coincidence*. Given sufficient time and space, even random or pseudo-random landscape processes are bound to produce analogous outcomes. Perhaps coincidence accounts for a few local features, but this is not sufficient to explain the large global number of similar landscape patterns.

Global (general) laws clearly play an important role. Though local place and history factors are vital in explaining and interpreting landscapes, repetition of certain features in geographically separated locations with different histories implies, and can often be directly linked to, one or more universal laws or principles. For example, the layered vertical structure of tropical rainforest vegetation despite large contrasts of the constituent plant can be explained by general principles of allocation and utilization of sunlight resources and processes of niche generation. Similarly, repetition of characteristic dune morphologies in sandy deserts around the world can be put down in part to the laws governing aeolian sand transport, as described above.

Another class of explanation for commonality of some forms is *overwhelming control*. That is, a certain process or factor is so overwhelmingly important that it overprints or obscures other factors. For instance, the forces of alpine glacial erosion are strong enough that, regardless of rock type, tectonic setting, or preexisting topography, certain common features (e.g., cirques, aretes, U-shaped valleys) are produced everywhere alpine glacial erosion occurs. This leads to a distinctive visible signature of glacially eroded terrain wherever it occurs (Figure 5.8). Similarly, where fluvial erosion is strong enough and lasts long enough, it overrides other factors and produces characteristic "badland" topography (Figure 5.9).

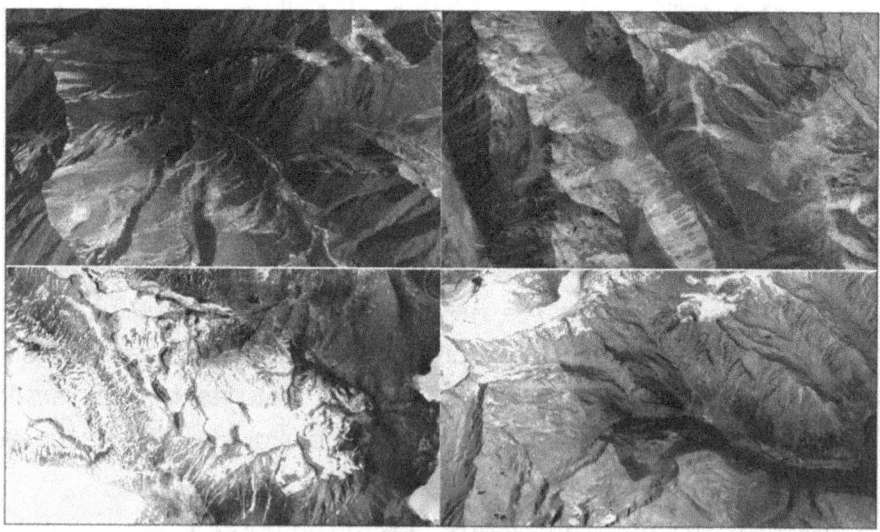

Figure 5.8. Glacially eroded topography illustrated by GoogleEarth™ images. Clockwise from upper left: near Qiongguo, Tibet; Mt. Cook region, New Zealand; near Motase Lake, western Canada; Maso Corto, Italy.

The concepts of evolution toward climax soils and vegetation communities or of characteristic landforms linked to climate or geological settings may be categorized as *geographical attractors*. These concepts recognize that different environmental settings will

Figure 5.9. Badlands topography. (A) Tsagaan Survarga, Mongolia; (B) Badlands National Park, South Dakota; (C) Grazing land, Western Australia; (D) Gover Gulch Badlands, Nevada. Photo credits: Arthouse Studio, Amaury Michaux; Government of Western Australia; Goodfreephotos.com.

have different outcomes of landscape evolution but postulate a single dominant attractor within geographical areas. The latter may be specified by location (e.g., equatorial Africa, Eurasian steppes) or generically (e.g., tropical savannas, alpine tundra). Geographical attractors thus explain repetition of landscape characteristics within a geographical zone and imply some degree of similarity be-tween similar zones (e.g., coniferous forests of Europe and North America and arid deserts of South America and Australia). Usu-ally underpinning geographical attractors, implicitly or explicitly, is some notion of overwhelming influence (for instance, that climate overpowers other influences to produce landforms, soils, or bio-geographic patterns) or of global laws (for instance, how chemical reactions involved in weathering or nutrient cycles are facilitated or constrained in different settings).

Goal functions, as described above, are often devised to ex-plain commonalities. For example, a wide variety of ecosystems seem to evolve or self-organize toward increasing energy through-put, inspiring the proposal that maximum power, energy cycling, emergy, or productivity is a goal function (see Table 5.1). Another example is the widespread occurrence of concave-up river longitu-dinal profiles, attributed by many to a goal function of balancing sediment supply and transport capacity (see below for another interpretation).

Emergence, in my view, accounts for whatever success goal function explanations and interpretations have. For example, given the slope gradient constraints on stream channels (minimum slope required to move water, maximum reach-scale slope possible for a given channel material) and assuming a requirement to move flow from headwater source to outlet, then a stream capable of eroding its bed will, given sufficient time, approach a concave profile. It is not necessary to invoke a goal function of balancing sediment sup-ply and transport capacity. And to the extent that increasing energy throughput confers some advantages for survival and replication of ecosystem entities, then a finite resource base and (a trend toward) biological saturation are all that is necessary to result in the trends predicted by maximum entropy production and related hypotheses. If similar phenomena can emerge from a few simple rules or con-straints, and if these are applicable in multiple locations, then we should expect them to emerge repeatedly.

The principle that ties many of the possible explanations above together is *selection*. More efficient flux pathways, structures, and patterns are *more likely* to persist and grow than less-efficient ones, be they preferential flow paths in soils, nutrient cycles in ecosystems, or surface channel systems. More durable (stable, resilient, resistant) forms are *more likely* to persist and be preserved than less-durable alternatives. And organisms that are more successful in surviving and reproducing are *more likely* to pass their genes to future generations and to occupy or create niches in ecosystems. The italics are there to emphasize that selection is probabilistic, not deterministic; chance, luck, and coincidence play a role. Chapter 6 is devoted to selection principles in landscape evolution.

Emergence and selection represent a consilience of views on the destinations of landscape evolution.

NOTES

1. G. Cooper, "Must there be a balance of nature?" *Biology and Philosophy* 16 (2001): 481–506.
2. G. K. Gilbert, *Report on the Geology of the Henry Mountains* (Washington: U.S. Geographical and Geological Survey, 1877); J. T. Hack, "Interpretation of Erosional Topography in Humid Temperate Regions," *American Journal of Science* 285 (1960): 80– 97. See also these commentaries: L. J. Bracken and J. Wainwright, "Geomorphological Equilibrium: Myth and Metaphor? *Transactions, Institute of British Geographers* 31 (2006): 167–178; L. J. Bracken and J. Wainwright, "Equilibrium in the Balance? Implications for Landscape Evolution from Dryland Environments," *Geological Society of London, Special Publications* 296 (2008): 29–46; G. C. Nanson and H. Q. Huang, "A Philosophy of Rivers: Equilibrium States, Channel Evolution, Teleomatic Change and the Least Action Principle," *Geomorphology* 302 (2018) 3–19; W. H. Renwick, "Equilibrium, Disequilibrium, and Nonequilibrium Landforms in the Landscape," *Geomorphology* 5 (1992): 265–276; C. E. Thorn and M. R. Welford, "The Equilibrium Concept in Geomorphology," *Annals of the Association of American Geographers* 84 (1994): 666–696.
3. I have referred to the utility of the steady-state soil thickness concept despite its flaws in describing actual soil or regolith evolution as a convenient fiction: J. D. Phillips, "The Convenient Fiction of Steady-State Soil Thickness," *Geoderma* 156 (2010): 389–398.

4. J. Wu and O. L. Loucks, "From balance of nature to Hierarchical Patch Dynamics: A Paradigm Shift in Ecology," *Quarterly Review of Biology* 70 (1995): 439–466.

5. M. A. Fonstad and W. A. Marcus, "Self-Organized Criticality in Riverbank Systems," *Annals of the Association of American Geographers* 93 (2003): 281–296.

6. P. Bak, *How Nature Works: The Science of Self-Organized Criticality* (New York: Springer, 1996); P. Bak, C. Tang, and K. Wiesenfeld, "Self- Organized Criticality: An Explanation of 1/f Noise," *Physical Review Letters* 59 (1987)L 381–384; S. Hergarten, *Self-Organized Criticality in Earth Systems*, 2nd ed. (Berlin: Springer, 2013).

7. J. Croke, R. Denham, C. Thompson, and J. Grove, "Evidence of Self-Organized Criticality in Riverbank Mass Failures: A Matter of Perspective?" *Earth Surface Processes and Landforms* 40 (2015): 953–964.

8. J. Wu, K. B. Jones, H. Li, and O. L. Loucks, eds., *Scaling and Uncertainty Analysis in Ecology* (Berlin: Springer, 2006).

9. M. Scheffer, S. H. Hosper, M. L. Meijer, and B. Moss, "Alternative Equilibria in Shallow Lakes," *Trends in Ecology and Evolution* 8 (1993): 275–279; M. Scheffer, E. H. van Nes, "Shallow Lakes Theory Revisited: Various Alternative Regimes Driven by Climate, Nutrients, Depth and Lake Size," *Hydrobiologia* 584 (2007): 455–466.

10. J. B. Thornes, "The Ecology of Erosion," *Geography* 70 (1985): 222–235. Examples of later work elucidating the processes and confirming Thornes's analysis include A. D. Abrahams, A. J. Parsons, and J. Wainwright, "Effects of Vegetation Change on Interrill Runoff and Erosion, Walnut Gulch, Arizona," *Geomorphology* 13 (1995): 37–48; A. J. Parsons, A. D. Abrahams, and J. Wainwright, "Responses of Interrill Runoff and Erosion Rates to Vegetation Change in Southern Arizona," *Geomorphology* 14 (1996): 311–317.

11. E. W. Lane, "The Importance of Fluvial Morphology in Hydraulic Engineering," *Proceedings of the American Society of Civil Engineers* 81 (1955): 1–17.

12. Examples of Liang's work in this area: L. Liang, "Beyond the Bioclimatic Law: Geographic Adaptation Patterns of Temperature Plant Phenology," *Progress in Physical Geography* 40 (2016): 811–834. L. Liang and M. D. Schwartz, "Testing a Growth Efficiency Hypothesis with Continental-Scale Phenological Variations of Common and Cloned Plants," *International Journal of Biometeorology* 58 (2014): 1789–1797.

13. R. C. Dewar, "Maximum Entropy Production and Plant Optimization Theories," *Philosophical Transactions of the Royal Society* B 365 (2010): 1429–1435; B. D. Fath, B. C. Patten, and J. S. Choi, "Complementarity of Ecological Goal Functions," *Journal of Theoretical Biology* 208 (2001): 493–506; S. E. Jrgensen, *Integration of Ecosystem Theories: A Pattern*, 3rd ed. (Dordrecht, the Netherlands: Kluwer, 2022); B. C. Patten, "Network Integration of Ecological Extremal Principles: Exergy, Emergy, Power, Ascendency, and Indirect Effects," *Ecological Modelling* 79 (1995): 75–84;

R. E. Ulanowicz, S. E. Jrgensen, and B. D. Fath, "Exergy, Information and Aggradation: An Ecosystems Reconciliation," *Ecological Modelling* 198 (2006): 520–524.

14. While I was writing this book, Axel Kleidon published an excellent overview of thermodynamics in the global Earth system and showed or reinforced the equivalency of several sometimes seemingly contradictory thermodynamic goal functions: A. Kleidon, "Working at the Limit: A review of Thermodynamics and Optimality in the Earth System," *Earth System Dynamics* 14 (2023): 861–896

15. The barchan example is based on: H. El Belrhiti and S. Douady, "Equilibrium Versus Disequilibrium of Barchan Dunes," *Geomorphology* 125 (2011): 558–568; C. Groh, I. Rehberg, and C. A. Kruell, "How Attractive Is a Barchan Dune? *New Journal of Physics* 11 (2009): https://doi.org/10.1088/1367-2630/11/2/023014; P. A. Hesp and K. Hastings, "Width, Height, and Slope Relationships and Aerodynamic Maintenance of Barchans," *Geomorphology* 22 (1998): 193–204; K. Kroy, S. Fischer, and B. Obermayer, "The Shape of Barchan Dunes," *Journal of Physics: Condensed Matter* 17 (2005): https://doi.org/10.1088/0953-8984/17/14/012; G. Sauermann, P. Rognon, A. Poliakov, and H. J. Herrmann, "The Shape of the Barchan Dunes of Southern Morocco," *Geomorphology* 36 (2000): 47–62.

6

Selection and Landscape Evolution

VARIETIES OF SELECTION

If you choose to eat a healthy diet, exercise, and avoid tobacco, those choices (selections) by no means guarantee that you will never fall victim to heart or lung disease. But they greatly increase the odds—the probability—of your survival and persistence. That is at least somewhat like how selection works in landscape evolution. Though the selection is not influenced by choices, except where humans are directly involved, various selection processes greatly increase the likelihood of some outcomes and decrease the likelihood of others. Thus, selection—probabilistically but not deterministically—moves evolution along certain trajectories.

Some scientists try to restrict discussions of selection and evolution to Darwinian natural selection and biological evolution, and some get their dander up when some of us do otherwise. To circumvent these objections (which came up, again, in reviews of the proposal for this book), I tried to find some appropriate synonyms. But get out your thesaurus and see what the existing options are; none of them capture what we are talking about here. And inventing new words would not really help my ability to communicate to you. And finally, objections aside, there is a long tradition in the Earth and environmental sciences of studies and discussions of

Mysterious Ways. Jonathan D. Phillips, Oxford University Press. © Oxford University Press (2025).
DOI: 10.1093/9780197755129.003.0006

evolution and selection that includes, but transcends, biological selection and evolution. So, for the bio-linguistic word police—get over it (please).

In this chapter, we'll indeed start with the most familiar form of selection, which is indeed Darwinian natural selection. We'll then discuss ecological selection and selection and evolution phenomena at the ecosystem scale. Selection also occurs in abiotic phenomena, and we'll review those ideas. Next is exploration of some examples of specific selection phenomena, such as preferential flows and resistance selection, that are particularly important in landscape evolution. Finally, to tie together the role of selection in landscapes where multiple selection processes (biotic, abiotic, and biogeophysical) operate simultaneously, three case studies are presented.

Natural Selection

Sometime in the late Pleistocene, maybe thirty or forty thousand years ago, some wolves were hanging around human camps looking for an easy meal. Wolves varied in their fight-or-flight response, and those that were less aggressive and friendlier to people were allowed, or even encouraged, to hang around, as people realized the benefits of a canine alarm system. Thus began the genetic divergence of *Canis lupus*. Continued "social selection" of the wolves more friendly and useful to humans—and later selective breeding for canines with particular talents, abilities, and traits—resulted in the emergence of (depending on who you ask) a new subspecies (*Canis lupus familiaris*) or a new species (*Canis familiaris*, the domestic dog). The American Kennel Club now recognizes more than 450 dog breeds, ranging from English mastiffs, many of which top out at >110 kg, to tiny little "purse dogs," where even healthy adults may weigh in at <1 kg. A comparably broad range of variation exists in other traits.

This is just one example of *artificial selection* (Figure 6.1). Charles Darwin's critical insight was that nature could also select individuals, though in a much slower, more haphazard, and unconscious manner. *Natural selection* was coined by Darwin to contrast the latter phenomenon with the selection practiced deliberately by humans. This is the basis for biological evolution. If you, as an individual organism, are better able to make use of the available resources, withstand the local stresses, and get your reproduction on, you are more likely to survive, thrive, and send your genes into the next generation. If the traits giving you those advantages are heritable, they may be passed on to the next generation. Thus, evolution.

Does biological evolution occur gradually and incrementally or in episodic bursts interspersed with long periods of stasis? Does it always result in increased complexity? What are the specific genetic mechanisms involved? All these questions, and more, are still under debate. But the fundamental notion of evolution by natural selection is firmly established, and the basic elements are not credibly challenged by biologists or Earth scientists.

Natural selection is probabilistic, not deterministic—"more likely" to survive and reproduce is a key element. Sometimes the best adapted, healthiest, most genetically advantaged individuals do not survive and reproduce. The better adapted and fitter are more likely to pass their genes to the next generation, and this is what happens on average, but not inevitably. Another element of luck occurs via chance genetic mutations and genetic drift. Most of these are unhelpful or disadvantageous, but a few are advantageous.

An additional note on natural selection: even when we talk about genes, the abiotic environment is part of the picture. Lekevicius shows convincingly that genotype diversity is maintained by both abiotic and biotic factors.

We'll discuss several different kinds of selection that don't directly involve humans, but I'll use *natural selection* specifically in reference to biological evolution and the Darwinian account of species evolution.

Figure 6.1. Another example of artificial selection: maize. Humans began to domesticate a grass, teosinte, about 9000 years ago in what is now southern Mexico. Culling individuals with undesirable traits and breeding plants with desirable ones, farmers transformed teosinte (*Zea mays parviglumis and Z. Mexicana*) into its nearly unrecognizable ancestor, *Z. mays mays*, or maize (often called corn in the United States). More than fifty commercially recognized varieties of maize now exist, deliberately selected for suitability for specific environments and uses and resistance to different stresses, pests, and pesticides. The number of genetically distinct varieties is even higher. Artificial selection by agricultural seed and chemical companies is now reducing genetic diversity of maize by way of a variety of marketing forces.

Public domain, https://commons.wikimedia.org/w/index.php?curid=183925.

Ecological Filtering

In biology and ecology, selection also operates at scales broader than individuals or species. Filtering occurs at ecological levels and is, straightforwardly enough, referred to as ecological filtering or *ecological selection*. The abiotic environment strongly affects, and sometimes dominates, ecological selection.

Organisms have minimum amounts of certain resources they must obtain to survive, with higher quantities required for good health and reproductive success. They also have tolerance limits for the kinds of stresses they must face—e.g., heat or cold, wetness or aridity, disturbances, as well as more optimal ranges that allow them to do more than merely endure (Figure 6.2). Biota are equipped by natural selection with varying abilities to adapt to environmental conditions, and the environment filters out non- or poorly adapted individuals by denying them resources and/or subjecting them to unhealthy or lethal stresses. Simple examples are arid environments that do not allow plants without suitable adaptations to water limitations to establish or thrive or wetlands that filter out plants that can't stand frequent soil saturation or inundation and the chemical conditions associated with it (such as anaerobiosis). Ecological filtering is by no means limited to abiotic factors, however. Sulfate-reducing bacteria, for instance, may filter out plants not adapted to high levels of hydrogen sulfide, and plants themselves may create shade conditions that favor shade-tolerant and inhibit shade-intolerant vegetation.

Though it is most obvious in harsh environments where extreme conditions or limitations require special adaptations, ecological filtering exists in all landscapes. In the southeastern United States, for example, karst areas with limestone parent material are very well suited to many plants and agricultural activities. However, the high pH of the soils is unfavorable for many of the plants common in the southeast, which are adapted to the acidic soils usually found in non-karst areas of that humid subtropical region.

Ecological selection is also influenced by reciprocal, coevolutionary effects. Through ecosystem engineering, biota modify

Figure 6.2. Ecological filtering. Perpetually wet conditions in this eastern North Carolina swamp exclude plants not adapted to wetness and favor plants such as these bald cypress trees (Taxodium distichum) which are well adapted to saturated soils and standing water.

their habitats, thereby changing the ecological selection pressures, both positively and negatively, for the engineer species and others. Feedbacks to natural selection may also occur. The principle that "ecological opportunity" associated with environmental factors may drive adaptive diversification has been prominent since Darwin.

More and more empirical evidence to support adaptive diversification has emerged recently. Syntheses of adaptive diversification and ecological opportunity studies led to a mechanistic conceptual framework describing the general phenomenon. Ecological opportunities can allow persistence of a genetic lineage and stimulate divergent natural selection. This depends on the availability of niches that allow a population with a new (i.e., previously absent from that population) phenotype to persist within that community. It also requires niche discordance, defined as diversifying selection arising from mismatches between a population's niche-related adaptive traits and the new environmental conditions (i.e., the ecological opportunity). Thus, environmental factors and their spatiotemporal structure, along with biological properties of a lineage that determine its capacity to diversify, control evolutionary responses to ecological opportunity.[1]

Coevolution and Natural Selection

Ecosystem engineering happens over ecological timescales. When this affects natural selection on longer timescales of biological evolution, it is called *niche construction*.

Niche construction essentially manufactures new ecological opportunities, which stimulate adaptive divergence. One example is the diversification of metazoa (multicell animals) from the Ediacaran to the Cambrian (that is, between about 635 and 539 Ma). The diversification was partly driven by positive ecological feedbacks that occurred mostly due to the appearance of ecosystem engineers that modified the environment, mainly via bioturbation and biochemical changes.[2] Another example is evidence that tetrapods (four-legged vertebrates) evolved in the Devonian (about 419 to 359 Ma). The appearance of trees in the Devonian is well known, and limbed critters with long necks exploited the ecological opportunity with their adaptations selected for scavenging and hunting in shallow-flooded woodlands and oxbow lakes. This happened between the evolutionary appearance of trees that grew in flooded or ponded conditions (itself likely due to ecological opportunities) and before predators appeared.[3] Evidence of coevolution of microbes and intertidal landforms dates to the Precambrian.[4]

For an example from the plant kingdom, consider the Cape Floristic Region of southern Africa. Here, dated plant phylogenies (evolutionary family trees) reflect Neogene (beginning about 23 Ma and ending with the start of the Quaternary) climate and landscape evolution. Geomorphic change led to a greater variety of substrates and therefore a new set of niches and adaptive radiation. Fenton (Woody) Cotterill, of the University of Stellenbosch, South Africa, and colleagues have produced many examples of the relationships between genomics, biological evolutionary change, and geological events, based mainly on evidence from southern Africa.[5]

Evidence of ecosystem engineering creating ecological opportunities or otherwise exerting selective pressure on organisms at ecological timescales is abundant in the literature on plant-soil interactions. The fundamental idea is that pedogenesis exerts selective effects on organisms that in turn influence pedogenesis. Ecological and evolutionary dynamics in plant-soil interactions are driven by feedbacks across levels of organization. Specific examples show how tree genotypes influence microbial communities, which in turn affect soil nutrients, which have selective influences on plant performance.[6]

Ecosystem Selection

As discussed in Chapter 2 on supra-organic concepts, selection can also operate at broader levels, including ecosystems. We'll briefly revisit that here.

The basic idea is this: when patterns or processes of resource utilization, such as biogeochemical cycling or water use by plants, provides some advantages for survival and reproduction to communities or networks (as well as individuals), selection favors those processes and patterns. As is the case with all selection, this is probabilistic—the advantage-conferring mechanisms and structures are more likely to survive and persist and to expand or to be replicated. This is all that is necessary to produce directional trends—evolution in ecosystems. Those directional trends are often evolutionary trajectories that, on average, maximize rates of resource use and recycling, energy and mass flux efficiency, and sometimes biodiversity.[7]

Ecosystem evolution toward certain optimal patterns of matter and energy use is evident in the work on ecological goal functions described in Chapter 5. On a longer timescale, evidence of evolution of ecosystems is present in the paleoecological record. This evidence applies to general categories of ecosystems (e.g., intertidal environments, subhumid grasslands) and to specific ecosystems.

Paleoenvironmental records show that ecosystems change over time; those changes are triggered by both external factors and internal dynamics. Climate and other environmental changes and disturbances may trigger ecosystem change and influence evolutionary trajectories, reflecting the interplay of species evolution, abiotic change, and biotic-abiotic interactions.

ABIOTIC SELECTION

Least Action Principle

Aristotle (384–322 BCE) famously said that "nature does nothing in vain." The ancient Greek scholar pointedly did not restrict his observation to life, and the principle indicates that nature maximizes efficiency by *selecting* the least-wasteful way of getting anything done. Perhaps the simplest and most general manifestation is the principle of mechanics known as the principle of least work or the least action principle (LAP), which we encountered in Chapter 5. In essence, the LAP holds that nature always finds the most efficient path (or in Aristotle's terms and later echoed by Isaac Newton, nature does nothing in vain). Aristotle's quote may be ascribed some metaphysical implications, but the LAP has the same implications, for specific situations in physics, with no necessity for teleology or for intentionality.

Envision a soil particle splashed away from the point of impact by a raindrop, a sand grain or seed blown by the wind, or any other convenient case of a moving object. For such a situation, the LAP can be defined by the average kinetic energy minus the average potential energy being as little as possible for the path of an object going from one point to another. This means that work (e.g., heat flux, water flow, sediment transport, vegetation productivity) in landscapes is accomplished with as little energy as possible. For the case of, say, a fired artillery shell or a thrown baseball where the necessary measurements are relatively simple, the path of motion dictated by the LAP is straightforwardly calculated and is deterministic. For most other work, particularly in environmental systems, the situation is more complicated.

While nature may indeed follow the most efficient path in physical mechanical systems, this is not selection in the sense of factors resulting in differential survival, preservation, and reproduction or propagation. It is patterns, structures, processes, and networks—landscape traits, if you will—that enable or inhibit such efficiency that are selected for or against.

A broader but related principle holds that organisms (including humans) will choose the path of least effort or resistance to achieve a goal—the principle of least effort. A rolling boulder or a thrown object has, in essence, perfect information. All it needs to "know" is there from the git-go: initial position, applied force, mass, and resistance. Not so for humans, who must choose a course of action, almost always without perfect information. Other organisms also lack complete and accurate information beyond their immediate situation. Organisms may also have multiple or variable goals. The optimal travel route through a landscape may vary depending on whether the purpose is pursuit, escape, foraging, A-to-B transportation, or for humans, relaxation, sightseeing, or exercise. Biota also often have multiple possibilities for pursuing goals. This is most obvious for humans. Even a single simple goal such as "maximize profits" or "get high" may involve multiple tactics, alone or in combination. And the projectile always reaches the location dictated by mechanics, but in other cases goals are not necessarily achieved.

The least action principle is applicable at the scale of process mechanics in landscapes. But since landscapes involve multiple entities, multiple goals, and incomplete information (for the elements within it and the researcher studying it), the least effort principle may be a better analog for landscapes.

He Qing Huang and Gerald Nanson led the way in directly connecting the LAP with selection in landscapes through their studies of fluvial systems. They argued in 2000 that stream channels

adjust in such a way as to maximize flow efficiency, thereby doing the work of moving water and transporting sediment with the
least amount of energy (Figure 6.3). Later, they also showed how
the LAP can explain anastamosing patterns in some rivers and
demonstrated how iterative adjustments work in stream channels
to achieve or approach maximum flow efficiency. By 2017, Nanson
and Huang had further expanded these ideas into a comprehensive theory of how the LAP drives self-adjustment in channels,
thus making the principle the primary control of form-process
interactions in many alluvial channels.[8]

Figure 6.3. Iterative adjustments to maximize flow efficiency are readily evident
in sand-bed rivers such as the Sabine, on the Louisiana-Texas border. The unconsolidated sand allows rapid adjustments changes in discharge, slope, and
other factors.

An overarching evolutionary approach to fluvial geomorphology
based on selection for maximum flow efficiency due to the LAP was
articulated by Nanson and Huang in 2018. In this, they followed
Luna Leopold (1915–2006) in drawing an explicit analogy between
Darwinian natural selection and geomorphic evolution. Leopold observed that rivers have similar forms without being identical and
with uniqueness of individual channels and responses to changes.
He also noted tendencies to develop forms that minimize energy
expenditure and to equalize it along a river's length. If the energy
expenditure criteria are equated to fitness, then the characteristics

indicated by Leopold indicate selection. Importantly, though, Nanson and Huang showed that evolution due to selection processes in fluvial channels can be derived independently, and directly from the energetics of fluvial processes, without resorting to any analogies with biological evolution.[9]

Nanson and Huang explicitly invoked teleology or teleomatics (as did Leopold) based on progression toward an end-state determined by a natural law: the LAP. I take some issue with this for semantic reasons—teleology is strongly associated with religious and creationist views by many scientists and laypersons rather than the more general principle of progression toward a predetermined end point meant by Nanson and Huang. The teleomatic perspective also downplays the possibility of multiple possible outcomes and emergence in geomorphic evolution, even though Huang and Nanson recognize that least action may be achieved in multiple ways. Efficiency selection probabilistically nudges geomorphic systems toward "fitness" but does not predetermine any particular outcome. While I have no substantive quarrel with their perspective, I prefer to stay away from teleo-isms. But heck, let's stay away from the tiresome parsing of words.

Robert Pascal and Addy Pross (University of Montpelier and Ben Gurion University) in 2016 proposed a logical connection between the physical and biological worlds based on stability. Stability has both time and energy facets, of which time is the most general. Time, expressed as persistence, led them to the persistence principle: nature seeks persistent forms. In 2017, they developed the idea in the more specific context of thermodynamics of chemical kinetics. Dynamic kinetic stability is based on time/persistence (rather than free energy) and provides a basis for understanding evolution. Pascal and Pross argue that kinetics is important in biological evolution and that in non-living systems all selection is thermodynamic. The same principles govern physical and biological systems, they assert.[10]

Not all selection in abiotic systems is directly linked to thermodynamics (e.g., see the resistance discussion below). However,

Pascal and Pross's work is at least an additional example that evolution of non-living systems (or the non-living components of systems) can be addressed independently of analogs with Darwinian
natural selection.

Geophysical Selection

Most fluvial systems are strongly influenced by biota, but the selection principles of Leopold, Huang, and Nanson are independent
of biology. Selection is also implied by multiple principles of purely
geophysical and geochemical processes.

Patterns of heat flux in the atmosphere self-organize via selection to maximize the rate of potential energy production (equivalent
to maximum entropy production, MEP), and selection operates on
ocean-atmosphere heat fluxes to minimize entropy exchange with
the external environment. Fluid convection in the atmosphere and
hydrosphere is governed by a selection principle of MEP, where
at steady state convection maximizes heat flux and thus entropy
export. Turbulent energy dissipation in fluid flows maximizes entropy export at steady state. Selection in these cases means that
of all possible flux patterns, the one(s) that achieve MEP (or some
relative thereof) are the ones that occur.[11]

General selection principles as they regard flows were formal-
ized by Adrian Bejan of Duke University as the constructal law: "for
a flow system to persist in time (to survive) it must evolve in such
a way that it provides easier and easier access to the currents that
flow through it." Phenomena that illustrate this type of development
include wedge-shaped turbulent shear layers, jets and plumes,
the frequency of vortex shedding, Bénard convection in fluids and
fluid-saturated porous media, and coalescence of solid parcels
suspended in a flow. Bejan also argued that global ocean and
atmospheric circulation reflects the constructal law. A direct, explicit link exists between evolution/selection and thermodynamics,
Bejan argues, such that thermodynamic systems (essentially,
all systems that move or change) evolve. Constructal-type

behavior arises from traditional path-of-least-resistance reasoning if selection results in persistence of the preferred flow paths.[12]

Where work is accomplished with the minimum amount of the available energy, then the remaining energy must be dissipated as entropy, so the LAP is consistent with, and in some cases equivalent to, MEP. Before (and since) Huang and Nanson linked the LAP and efficiency selection to fluvial morphology, others also argued that fluvial channels adjust to maximize efficiency, though this was phrased in various ways and based on various criteria. Selection principles have also been applied to channel networks and drainage basins by Axel Kleidon (Max Planck Institute) and colleagues.[13]

Selection also operates in groundwater flow systems to maximize fluxes, according to Canadian hydrogeologist Stephen Worthington. Lag times are lower and flow velocities and celerities (a measure of how quickly a wave passes through a medium) higher in dual-porosity systems (matrix plus preferential flow paths) than in single-porosity systems. Selection moves the system toward dual porosity, where the matrix provides most storage, with preferential flow paths providing most transport.

Speaking of which. . . .

PREFERENTIAL FLOW

You've heard it and probably said it: water follows the path of least resistance. This reflects the intuitive understanding that water flows are subject to selection. The saying is also sometimes given as water *finds* the path of least resistance, but as H_2O cannot want, seek, or find anything, a less-compact but more accurate way of putting it is that the path of least resistance is usually the most efficient route for water transport, and therefore such paths are more likely to occur and persist than less-efficient alternatives.

Water flows tend to be concentrated rather than dispersed or uniform (that is, in threads or channels rather than sheets), and those concentrated flows are more likely to occur along paths that are more efficient due to being steeper, deeper, or less obstructed. In hydrology, this is referred to as *preferential flow*, a term that seems to have originated in soil hydrology to distinguish it from water moving through soil as a more-or-less homogeneous mass. However, preferential flow is the rule, not the exception, in all hydrological processes, at all scales, in all environments. Preferential flow is one of the most common expressions of selection in landscapes.

Surface Flow

Sheet flow, the opposite of preferential flow for surface runoff, does occur. But sheet flow is far rarer than concentrated flow, and when it does occur it is short-lived and spatially restricted. Preferential flow, by contrast, is everywhere, from rivers and streams to gullies and rills, to ephemeral threads of runoff, to those rivulets that form when it is raining on your windshield or windowpane. But why is sheet flow rare and transient, and why does runoff become concentrated and sometimes form channels?

Because selection.

Basic physics tells us that the velocity of flow (and we assume here that velocity indicates how efficiently mass is being transported) is a function of gravity, the steepness of the gradient along which it flows, the depth of the flow, and roughness or frictional resistance. As long as we're on Earth, gravity is constant. The steeper the gradient and deeper the water, the greater the velocity; and the higher the resistance, the slower the flow.

The slope, in terms of the physics involved, is the energy grade slope or the gradient of hydraulic head, but this is closely related to and ultimately constrained by the topographic slope of the surface. So, for a given mass of water moving down a given gradient, higher velocities can be achieved by concentrating the water to achieve greater depths rather than spreading it out, and by flowing around obstructions or areas of greater resistance. Depth also reduces resistance by drowning roughness elements—think of a pebble the size of double-aught buckshot, which is about 8.4 mm in diameter. In a flow that is 5 or 10 mm deep, this is a major obstruction. In a flow that is 100 mm (<4 inches) deep, not so much. And as the water gets deeper, the roughness element has less effect.

Physics tells us what knew intuitively already: concentrated flow is more efficient than diffuse flow.

Now let's think about the force exerted by the flow against the material it is flowing over or through. Again, good ol' Newtonian physics tells us this is determined by the specific gravity of water (itself determined by the gravity constant and the density of water, which for these purposes is so close to being uniform it can be treated as a constant), the depth, and the slope. Concentrated flows will be deeper than sheet or diffuse runoff, and the selected routes will also generally be the steepest available. So, the concentrated fluxes that maximize flow efficiency also maximize this force (mean boundary shear stress), making it more likely that the force exerted by the flow will exceed the resistance of the material, allowing erosion—of a channel.

When a channel forms, positive feedback really comes into play. The surface groove creates local gradients that draw more water to it, which tends to further increase velocity and shear stress (up to the point where other factors, such as the supply of water, become limiting). Positive feedback also happens because when channels or concentrated flows merge, the combined flow is more efficient than the individual ones. Next time it rains, watch raindrops sliding down window glass. When two drops come into contact, they merge, largely due to the nature of water molecules and surface tension. When two rivulets on the glass bump into each other, they also merge, and the converging flow is more efficient than the individual paths. Channelized flows thereby typically form branching, convergent networks. Where geologic and other constraints permit it, these branching networks generally form dendritic (tree-like) patterns. Dendritic networks are the most efficient way to collect and move fluids on the land surface (and in many other contexts) and are therefore subject to efficiency selection (see the *Net Benefits* section below).

Soil Moisture and Groundwater

Even more so than surface runoff, underground flow routes often opportunistically exploit more accessible, smoother (less-resistant) pathways associated with joints, fractures, and bedding planes in rock; cracks or textural boundaries in soil; plant roots and channels or macropores associated with decayed roots; faunal burrows and tunnels; pockets or layers of more permeable material; and probably other things I haven't noticed or thought of. Once these paths are carrying flow, they may be enlarged or enhanced by positive feedbacks (although they can also be obstructed or plugged by transported material).

However, preferential flow paths can also develop in homogeneous, or at least apparently homogeneous, material absent of the opportunistic routes mentioned above. This is most often observed in the phenomena of unstable wetting fronts and fingered flow. I first became aware of this in the 1990s via the work of Dutch hydrologists Coen Ritsema and Louis Dekker, who documented preferential flow paths in uniform sand and showed how it could create distinctive (if often short-lived, as sand structures tend to be) features in sand dunes and other sandy environments. Meanwhile, at Cornell, Yaping Liu, Tammo Steenhuis, and J-Yves Parlange were producing some of the seminal studies on unstable wetting fronts.[14]

The crux of it is this: Small, and often *very* small, variations in wetting create a locally favorable path for moisture flux because hydraulic conductivity is greater in wetter soil. Once these locally wetter spots are formed, they are enhanced by positive feedbacks, whereby the wet get wetter and the drier stay drier. This leads to dynamically unstable wetting fronts and fingered flow. This is a classic example of deterministic chaos. Small variations in initial conditions, undetectable by normal methods of soil and hydrological analysis, persist and become magnified over time.

Preferential flow leads to development of what is often called "dual porosity" systems consisting of rapid flow paths that can move water quickly, coupled with (in soils and groundwater) a matrix that stores water and allows only much slower flows. The evolution of dual porosity "store-and-pour" systems is explored in detail in Chapter 7.

In karst areas, water "finds" whatever openings or weaknesses are in the rock. The water gradually dissolves those passages, allowing more water in, and so forth. If the water supply is sufficient, the enlargement reaches the point where the rate at which the chemical reactions (dissolution of calcium carbonate and other soluble minerals) can occur rather than the amount of water becomes limiting—a threshold called a breakthrough event. Worthington has produced some of the clearest accounts of how these dynamics affect ground and surface water hydrology and the development of caves and other karst landforms. Swiss engineering geologist Marco Filliponi and colleagues have also done interesting work linking these preferential flow phenomena to the evolution of conduit systems and caves as and have also applied the principles to practical problems of tunnel construction and mining.[15, 16]

While links between preferential flow paths and landscape evolution are most evident in the development of fluvial channel networks and karst systems, even those associated with wetting front instability and fingered flow often persist and are reflected in soil morphology and associated with specific pedogenic features (Figure 6.4). For example, flow fingers tend to recur in the same locations even in water-repellent sandy soils and can even be associated with vertically oriented soil horizons. Flow finger persistence and recurrence has also been demonstrated in other soils. For my part, along with some of my students at East Carolina University, we linked wetting front instability with spatial variability of soil depth and thickness.[17]

Figure 6.4. Visual evidence of recurrent preferential flow in two soils in tropical Queensland, Australia.

Preferential flow and its relationship to structures and morphology of soils and other flow systems is discussed in detail in Chapter 7.

Not surprisingly, relationships also exist between preferential flow and irregular wetting patterns and microbial ecology. Shane Franklin (University of Delaware) and coworkers found that preferential flow paths are microbial hotspots, reflected in (among other things) local oxygen depletions. They characterized preferential flow paths and their associated microbes as dynamical biogeochemical entities. Feedbacks may also occur. Biting Li of Clemson University and colleagues found that nonlinear dynamics of infiltration lead to preferential flow, affecting water distribution in soil. In turn, preferential flow is affected by interactions among water, soil, plants, and microorganisms. Microbial exudates (stuff oozed out by the microbes) increased water holding capacity and thereby affected infiltration processes.[18]

EFFICIENCY SELECTION

A simple, though extraordinarily broad, principle applies to all the forms of selection treated here: "better" things differentially and preferentially arise, persist, grow, and replicate. The "things" are forms, processes, traits, patterns, networks, relationships, and individuals. Better might be more durable or stable, better suited to or adapted to environmental conditions, or more efficient. I use *efficiency selection* as an umbrella term because efficiency is not only advantageous per se but also contributes to durability, stability, adaptability, and fitness.

Efficiency can be defined in terms of accomplishing work (e.g., transporting mass, converting sunlight) using the minimum amount of available energy (consistent with the LAP) with respect to fluxes and storages of energy and mass. Efficiency can also be defined in terms of maximizing economy—for instance, the most efficient way to enclose a given volume (minimizing the ratio of surface area to volume) is a sphere. Physically, a sphere is the most stable three-dimensional shape, an example of how efficiency may relate directly to durability and stability.

Resistance also affects and often determines durability and may be measured in many ways depending on the situation. For instance, resistance to stress in organisms might be assessed in terms of phenotypic plasticity or tolerance limits. Resistance to erosion and weathering can be indicated by solubility, mineral stability, hardness, cohesion, or shear or compressive strength, for example.

Natural selection, for which efficiency or "better" could be based on criteria of adaptation, reproductive success, or environmental fitness, has been so exhaustively examined (and debated) that there's no need for me to say more. Some other examples of efficiency selection are outlined below.

Left Behind

When the going gets tough, the tough stay put.

That's the idea of *resistance selection*. Stress and mortality preferentially remove weaker organisms, leaving the more resistant organisms. Weathering and erosion rapidly remove weaker, less-resistant, less-stable, and more exposed materials and features, thus preferentially preserving more resistant and stable ones. Positive feedback can accelerate resistance selection via gradient selection (see below) as mass and energy fluxes become concentrated along channels and other pathways formed in weaker materials.

As in other types of selection, this is not necessarily a foregone outcome. Resistance selection does not *always* lead to persistence of more resistance morphologies and materials, partly due to disturbances and changes in boundary conditions. The relative preservation of fire-resistant vs. fire-vulnerable vegetation, for instance, can change as climate or land use modifies fire regimes. Organisms well suited for anaerobic environments lose their resistance advantage if those environments dry out. The same forms or material typically exhibit differential resistance to different processes and in different contexts. Carbonate rocks such as limestone, for instance, have low resistance to dissolution and chemical weathering but often have high physical strength. Quartz sand is low in solubility, strong, stable, and highly resistant to chemical weathering. But sandy sediments and soils lack cohesion and moisture-holding capabilities and may be highly susceptible to wind and water erosion, depending on topography, vegetation cover, and moisture regimes. Organisms often have substantial resistance to some pests, pathogens, or disturbance agents but are highly vulnerable to others.

Resistance changes, too. Chemical weathering of rock, for instance, reduces the physical strength of rock masses as the less-resistant components are removed. However, the durable leftovers and secondary minerals are often strongly resistant to further chemical weathering. Undermining of hillslopes reduces their resistance to failures and mass movements, while the resulting failures may result in more stable slopes. Aging and senescence of organisms decreases resistance, but decay of the least resistant components of organic matter (e.g., sugars, soft tissues) increases resistance of the organic matter *en masse* as more durable remnants (e.g., lignin, cellulose, bone) make up a greater proportion.

Finally, keep in mind that that forces of destruction or modification are often applied to heterogeneous materials. This is perhaps easiest to visualize in terms of denudation in layered rocks, sediments, and soils (Figure 6.5). In flysch terranes, for example, the forces of weathering and erosion encounter layers of sandstones, shales, and other lithologies that vary dramatically in resistance.

Figure 6.5. Differentially weathered sandstones and shales on Big Walker Mountain, Virginia. Exposure is about 1.5 m thick.

Some evolutionary biologists have applied resistance to natural selection. W. Ford Doolittle of Dalhousie University (Canada) identified selection by survival as an important component of evolution, arguing that differential survival of non-competing, non-reproducing individuals will also result in increasing frequencies of survival-promoting traits or adaptations. *Stability-based sorting* is the term used by Jan Toman and Jaroslav Flegr (Charles University, Prague) to describe resistance/persistence dynamics in biological evolution, and they linked stability-based sorting to broader selection principles of the sort discussed here. This theme has been echoed by Exeter University's Tim Lenton at the system level, calling it sequential selection. Ecological systems that destabilize their environment don't last long, he and his colleagues argued, as extinctions and reorganizations occur until some new stable (more resistant) state is achieved.[19]

Go with the Flow

The most efficient flux gradients are preferentially utilized, pre-served, enhanced, and replicated. That is the principle of *gradient selection*. The most efficient potential flow paths are preferentially selected—for example, a tumbling boulder or flowing water are more likely to follow the most favorable path. Second, use of or flow along these paths tends to reinforce their efficiency and contributes to their preservation. The principle is called gradient selection be-cause variations in potential in space (i.e., gradients) drive mass fluxes. Think of gravity-driven flows, the pressure gradients that drive wind and air movements, and chemical osmosis. The most efficient (fastest) path is (other things being equal) the one with the steepest gradient.

Positive feedbacks characterize the most persistent forms of gradient selection, where flux along the gradient enhances its flow efficiency. Examples include concentrated surface runoff incising a channel, flow along a karst fissure enlarging it by dissolution, and repeated use of a route by animals creating a trail. In other cases, persistence can be enhanced in a weaker way if use of or flow along a gradient just doesn't degrade its efficiency. Note that degradation can happen—think about too much traffic on an unpaved road mak-ing it muddy and rutted or about material transported by subsurface flow clogging pores or conduits.

Net Benefits

Michael Woldenberg (SUNY-Buffalo) was a pioneer in studying flow networks. In 1969, he showed that stream channel systems are organized in such a way that they balance two opposing tendencies: minimization of overland work of small streams in small watersheds and maximizing work savings in large channels. Michael Kirkby of Leeds University, another giant of geomorphology and hydrology, independently came to similar conclusions shortly after. Consistent with the LAP and the Second Law of Thermodynamics, if work is accomplished with the minimum amount of energy, then the leftover energy, called entropy and in stream channels dissipated in the form of friction and turbulence, must be maximized. The best way to do this for any system collecting fluids from an area and delivering it to a point (such as a watershed) is a branching, dendritic network. This also applies to systems that take fluid from a central point and distribute throughout an area or to a system that does both, such as a cardiovascular network. Ignacio Rodriguez-Iturbe of Texas A&M later came up with an interesting metaphor to illustrate this. Imagine a series of nodes (runoff production sites) spread across the area to be drained. A purely "socialist" network, providing the least overall flow distance, would route flow from one to the other all the way to the outlet, producing a single linear path winding through the drainage area. A purely "capitalist" network, minimizing distance from each node to the outlet, would be a series of N-1 (where N is the number of nodes) straight lines from each node to the outlet node. A branching dendritic network with topological characteristics of those found in nature is a compromise between pure socialism and hyper-capitalism.[20]

This basic tendency shows up in nature, where channel networks free to develop more or less independently of geological (or human) constraints on where channels can go invariably form dendritic networks. It also shows up in networks of blood vessels and other biological networks—including, of course, trees. Woldenberg, a geographer and geoscientist, followed up his work on channel networks by collaborating with biomedical scientists to show how the principle applies in biology and medicine. The principle has also been reproduced in laboratory experiments, computer simulation models, and demonstrated mathematically.

We've already seen how selection favors concentrated, channel-ized flows, and the merger of preferential flows when they intersect. The tendency of these to develop into branching dendritic networks can also be explained based on efficiency selection.

So why aren't all hydrological flow networks that have had suf-ficient time to develop dendritic? Keep in mind that sometimes geological factors either constrain or determine where channels can form. For instance, networks of karst conduits are strongly influenced by geological controls and only occasionally assume a dendritic pattern. The efficiency rewards of dendritic pattens may also conflict with other types of selection, and local selec-tion (described later in this chapter) may not always contribute to overall efficiency of whole networks. Gradient and resistance se-lection manifest themselves in the immediate vicinity of flow, and the water cannot "see" where it should go to create and maintain a maximum-efficiency network.

When branching dendritic networks are available for hydrological systems and many biological systems, however, they are advanta-geous and are therefore preferentially formed, preserved, and up to some limits, expanded. Thus, we have *network selection*.

Eco-Thermodynamics

Thermodynamic sounds as though it should be in the abiotic section. However, it has been discussed mainly (often without invoking selection explicitly) in ecosystems.

As discussed in Chapter 5, several optimal or extremal principles based on thermodynamics have been proposed in ecology. Many of these can be unified, at least on a conceptual or mathematical level. All at least imply selection if they do not explicitly mention it. Absent a guiding hand, or intentionality on the part of organisms or ecological systems (which they also do not mention), selection is necessary for these trends to emerge.

Alfred J. Lotka (1880–1949) was one of the earliest to explore these ideas. In 1922, he wrote, "The first effect of natural selection thus operating on competing species will be to give relative preponderance (in number or mass) to those most efficient in guiding available energy . . . . Primarily the *path* of the energy flux through the system will be affected." Lotka noted that where there is extra energy to be had, total energy flux can be increased by organisms able to exploit it. If energy supply is limited, available power can be increased via faster rates of turnover (because power is energy expenditure per unit time). If energy and other resources are not limited, he reasoned, selection tends to increase total mass and the rate of matter and energy circulation. His ideas are therefore linked not only to thermodynamics but also to biogeochemical selection (see below).[21]

Brian Fath of Towson State University found that in ecological succession, deviation from thermodynamic equilibrium and increasing organization are attributable to the maximization of useful energy and energy storage. This implies that succession selects for these factors. Henry Lin (1965–2019) studied ecosystem development at seven sites in North America, considering soils and biota. Progressive (vs. regressive) ecosystem development requires maximum entropy production, he found. Selection for maximum entropy production (implying maximum efficiency in performing work) has also been found in a variety of other landscape phenomena.[22]

SINGERS, SONGS, AND CYCLES

Traditionally, many scientists had trouble reconciling Gaian ideas about biosphere evolution, as well as ecosystem evolution, due to the entrenched opinion that selection can only apply to individuals. At this point, you know that's not the case, but a couple of decades ago there was a lot of anxiety about reconciling Gaia and Darwin. In stepped Andrei Lapenis (whom we encountered earlier). He synthesized several Russian ideas on the biosphere, evolution, and ecology from the early twentieth century to support the notion that more rapid (efficient) rates of energy and elemental cycling are preferentially formed and enhanced as ecosystems (and bio-spheres) evolve. While the advantages of faster cycling can accrue to individuals, they also benefit ecological systems as a whole. The more rapidly nutrients and water are used within a landscape system, the more different entities can use them. This idea, which he termed *biogeochemical selection*, means that if more efficient material use and faster cycling confers an advantage, that is all that is necessary to explain evolutionary trends toward greater efficiency as reflected in thermodynamic selection (and in Gaia theory).

This form of selection was explored in greater detail with respect to water and plants by Peter Eagleson, who linked biogeochemical selection to natural selection. Leckevicius (2006) also made the connection between biogeochemical and natural selection, noting that while individuals are the focus of the latter, ecosystems cannot exist without nutrient cycles, and ecosystems determine the direction of evolution. Nutrient cycling is an emergent property of the minimum ecological processes necessary to support carbon-based life. The phytotarium concept of William Verboom and John Pate involves biogeochemical selection, whereby specific plant layers and associated microbes create niches to maximize access to limiting resources of water and nutrients.[23]

Doolittle took this a step further. He proposed, using microbes as examples, that biogeochemical cycles are the unit of selection, combining reproduction with "re-production." His chosen metaphor was singers (individuals) and songs (cycles).

SELECTION IS *LOCAL*

Natural selection doesn't care about your lineage—or that of any organism. Whether working to advantage or weed out an individual, selection takes no account of how it will affect the lineage or a species. Individuals may be helped or hindered (or unaffected) by whatever traits may differ from the norm, and their likelihood of survival and reproduction is modified accordingly. The outcome has knock-on effects, tiny and incremental though they may be, to the lineage and the species, but natural selection is highly localized to individuals.

All selection, at least in terms of its mechanisms, is local. The accumulated impacts of local selection construct the overall pattern of, for instance, ecosystem interaction pattens, soil morphology, or hydrological networks, analogous to the way cumulative impacts of natural selection on individuals affects the evolutionary trajectory of a lineage.

As mentioned above, however, in some cases what is most optimal or efficient locally does not necessarily contribute to efficiency at broader scales.

The tip of a growing root, a tunneling insect, infiltrating water, and a boulder rolling or creeping down a hillslope can't detect the optimal path to achieve maximum efficiency of the root or tunnel network or the fastest path to the saturated zone or bottom of the hill (Figure 6.6).[24] At a given moment, each is affected only by its immediate environment—the gradients, obstacles, and resistance exactly there and then. Selection is not only local but sometimes downright myopic. This may be the case even when selection occurs deliberately, as when root growth seeks to optimize access to water and nutrients or when an animal seeks the easiest path through thick brush. The mechanisms for route selection are imperfect, and the "decisions" are highly local—the animals, roots, rocks, and water cannot "see" beyond their immediate surroundings.

Often, of course, the cumulative effects of local selection nudge the landscape towards maximum efficiency (or even achieve it). But the local nature of selection is one reason that landscapes aren't always characterized by maximum efficiency (or progress toward it).

I began thinking about the locality of selection in trying to solve a field problem in a limestone bedrock stream in central Kentucky. Along Shawnee Run, there was a paleochannel that had been abandoned at some point. Normally where this happens, abandonment is in favor of a steeper or more direct path. In this case, it

Figure 6.6. These sycamore (*Platanus occidentalis*) roots cannot detect the overall most efficient pattern for growing in their limestone substrate; they can only "see" the situation at growing root tips.

was the opposite—a shorter, steeper-sloped, more efficient channel had been forsaken for a longer, lower-gradient, less-efficient path. What the heck?

Tasnuba Jerin (Missouri State University) and I investigated this more deeply in 2017. We found that abandonment of the more efficient surface channel occurred due to capture of streamflow by a subsurface karst flow path. Though the old channel was the most efficient path to move water toward the fluviokarst system's Kentucky River base level, when a vertical conduit (shaft or swallet) opened (this is common in that region), going straight down this hole into a groundwater conduit and cave passage was *locally* more efficient than continuing down the channel. That underground section of the stream was later exposed by erosion (Figure 6.7).[25]

As often happens, once a phenomenon gets on your radar screen you begin to see it more commonly, both in the field and the literature. For example, one study published in 2018 showed

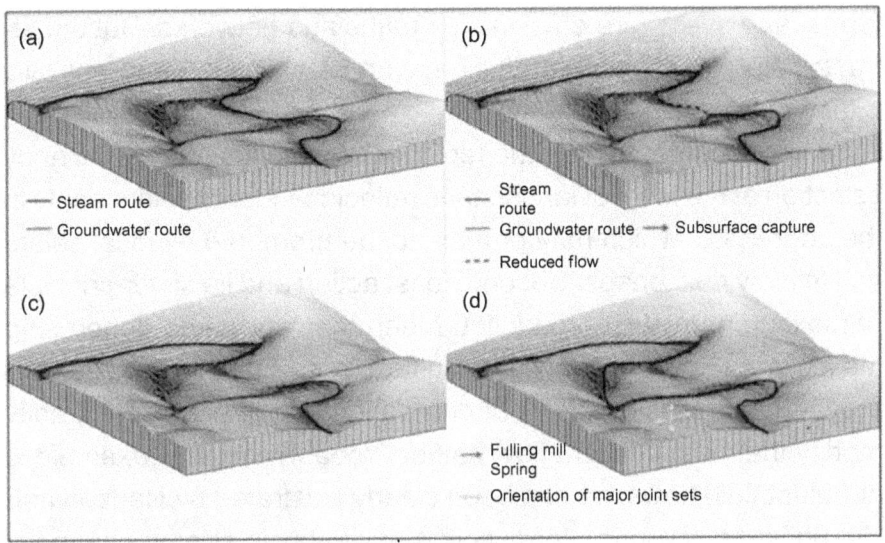

Figure 6.7. Reconstructed evolution of Icicle Bend on Shawnee Run, Kentucky, overlaid on contemporary topography. T. Jerin and J. D. Phillips, "Local Efficiency in Fluvial Systems: Lessons from Icicle Bend," Geomorphology 282 (2017): 119–130, Figure 9; used with permission).

that in a late-Cenozoic capture of a paleo Appalachian River, a new local opportunity created by erosion cutting through a resistant capstone rerouted the ancestral river. This resulted in the distance to the river's base level increasing by about 1000 km—another example of local efficiency overriding the broader scale. In a more general study of the linkages between hydrological fluxes, vegetation growth, and soil formation (which he continues to actively pursue), Allen Hunt of Wright State University showed how local efficiency may work against global (system-wide) efficiency using the technique of critical path analysis.[26]

A PERSISTENT PATH TOWARD INCREASED EFFICIENCY?

Not necessarily.

Efficiency selection and its variants suggest that, in general, landscapes should evolve toward prevalence of resistant, stable forms, and ever more efficient flux paths and networks. But the attractor state of maximum efficiency and stability is often not fully attained. Why?

First, of course, there's the fact that selection is local. Efficiency selection at the level of landscapes reflects tendencies that apply in the aggregate, which may or may not be promoted by local selection in individual cases. Second, least action and its sister principle of maximum entropy production mean that work gets done using the least amount of energy. That in turn means that any excess energy must be dissipated. That dissipation can directly or indirectly modify the most efficient flow paths. Probably the best example is turbulent fluid flows, as has been clearly illustrated by Nanson and Huang in their work on least action principles in stream channels. Energy dissipation in channels can create morphological variations and features that effectively dissipate available energy but work against maximum flow and transport rates (Figure 6.8).

Figure 6.8. Bedforms in this central Texas stream effectively dissipate energy at higher flows but work against maximum flow and transport rates.

Another factor that may impair steady progress toward maximum efficiency is canalization (see Chapter 4). Local selection can form paths that deviate from those promoting efficiency at the broader scale, as discussed above. Once this occurs, potential future paths are affected. The inherent dynamism of Earth and its environmental systems is a fourth barrier to attainment of maximum efficiency in landscapes. Everything changes, pretty much all the time, though sometimes gradually enough so that humans don't notice: boundary conditions, energy and mass inputs, and disturbance regimes are chronically altered by climate change, tectonic and volcanic activity, and sea level rise and fall (and, of course, humans). Independently of these external factors, landscapes themselves are changing due to their internal dynamics. Therefore, even as a landscape moves toward increasing efficiency in its current environmental context, that context is changing, and the target may be moving faster than efficiency selection can keep up with it.

Let's look at some additional complications to the landscapes-inexorably-get-more-efficient trend.

Multiple Simultaneous Selection

Even the smallest and simplest landscapes are full of stuff—different organisms, materials, forms, processes, flow paths, interrelationships. Selection happens simultaneously on many of them at any given moment. The efficiency criteria are different for various components, and increasing efficiency for one may not help (and may even impede or prevent) progress toward efficiency for another. Dam building by humans or beavers may make for better resource procurement, safety, and convenience for the dam builders but hardly enhances the efficiency of stream flow (Figure 6.9).

Paths taken by growing roots or burrowing fauna to maximize efficiency for their own needs will not necessarily be the most efficient paths for the soil moisture fluxes that follow these routes.

Figure 6.9. Beaver dam on the Smilga River, Lithuania.
Credit: Jūs ų Darbas (Creative Commons License 4.0).

Contrasting trends also occur in biogeochemical selection. In a study of temperate forests, for instance, it was found that rooting strategies to optimize the carbon budget differ from those optimizing water use. I spent a lot of time studying fluviokarst systems (Icicle Bend is one example), where flows will always prefer the most efficient route locally, but the contest between karst groundwater and surface fluvial flows can result in either or both being suboptimal.

Sometimes the dominant controls in landscapes can switch. This not only changes the ground rules for efficiency selection but can change the priority of selection processes within a landscape. Dov Corenblit (Université Clermont Auvergne) and coworkers pioneered the idea of biogeomorphic succession starting in 2007, which illustrates such switches in this and many other cases arising from landform-biota-soil-hydrology interactions. Biogeomorphic succession in fluvial systems, where Corenblit and company developed the concept, involves progress through four stages: (1) initial stage dominated by abiotic geomorphic controls; (2) a stage where plants become established, and mutual interactions among water flow, substrates, landforms, and vegetation become important; (3) ecohydrological and biogeomorphological feedbacks become strong and dominant; and (4) geophysical changes are minimized, and ecological processes and controls are dominant. Biogeomorphic succession indicates that while ecological and geophysical phenomena and their associated selection process(es) are present throughout, their relative importance changes through the sequence. Biogeomorphic feedback windows where such feedbacks prevail over either biotic or abiotic factors alone occur in some settings.[27]

A longer-term example comes from Annegret Larsen (Wageningen University) and colleagues, in their study of late-Quaternary evolution of a river in the tropics of northern Australia. Fire regimes exert important controls over the distribution of monsoonal rainforests there, but knickpoint migration can change the main controls. As knickpoints propagate upstream, the river downcutting dries out riparian zones by lowering water tables, degrading rainforests—putting this hydrogeomorphic phenomenon ahead of fire regimes as the major control. From the fluvial perspective, they found that river aggradation periods were controlled primarily by vegetation cover, while erosional episodes were dominated by abiotic factors.[28]

Biogeomorphic succession has also been applied to glacial forefields, lateral moraine slopes, subalpine alluvial fans, and coastal barrier islands. Working on a forested hillslope, my colleagues and I found evidence of biogeomorphic state transitions more complex than the four-part successional sequence. All these studies imply selection occurring in multiple components of the landscape and the likelihood of changes in the dominant selection factors.[29]

To summarize, while efficiency selection consistently operates in landscapes, we do not always observe a progression toward greater efficiency. This is due to the probabilistic nature of selection and the fact that the context for selection is often variable over time. Selection is local, and so its effects at specific locations do not always promote landscape-scale efficiency. Further, energy dissipation in one process or landscape sector may decrease efficiency in another. Selection operates on multiple components simultaneously, with effects that may counteract each other. Finally, the dominant controls of landscape evolution can vary over time, with the most important selection phenomena changing too.

OF RAZORS AND EMERGENCE

Selection is *emergent*. That means that it can happen without the direction of a guiding hand, without intentionality of the part of the selectee, and without invoking or postulating any unknown mechanisms or traits of the selection agents. Emergence is a simpler explanation than the alternatives, and it works.

To William of Ockham (1287–1347), an English philosopher and theologian, is attributed the principle of parsimony: *entia non sunt multiplicanda praeter nesessitatem*, or "entities must not be multiplied beyond necessity." This is a philosophical razor, which is a principle for "shaving off" unlikely or unnecessary explanations. Now referred to as Ockham's (or Occam's) razor, it is often rendered as: simpler explanations are more likely to be correct. Occam's razor means that unnecessary or improbable assumptions should be avoided and dictates that where multiple explanations are equally plausible or effective, the simplest one is likely to be correct and should be preferred.

Ockham's razor says that selection is emergent. If a trait enhances survival and reproduction and is heritable, it is more likely to be passed on to subsequent generations. That's emergence. If an entity is more stable, durable, or resistant, it is more likely to persist. If efficiency—of mass and energy flux, biogeochemical cycling, etc.—facilitates persistence (and perhaps growth and replication) and provides an advantage, it is more likely to occur, be maintained, and expand. All these explanations are emergent. They are not absolutely dictated by any law and require no intentionality internal or external to the selected phenomena.

Discussing selection beyond natural selection and as an emergent phenomenon brings us in contact with a school of thought known as *generalized Darwinism*. This holds, in essence, that the cornerstones of Darwinian biological evolution—variation, selection, and preservation and replication—are applicable to, with only slight exaggeration, darn near anything, as reflected in the fact that it is sometimes called *universal Darwinism*. Generalized Darwinism is most actively debated in the subfields of evolutionary psychology and evolutionary economics, but it can be detected, with or without the GD label, in many fields.

Most critiques of generalized Darwinism (and they are not hard to find) fall into two categories: (1) a lack of fidelity to biological evolution; or (2) an inability to solve every problem in evolutionary economics, system theory, etc. Such criticisms are not valid, though they are accurate. What about *generalized* do you not get? It clearly implies applicability beyond biological evolution, and you already know that we can reject claims that selection and evolution cannot be rigorously or properly applied outside biology. With respect to the "it doesn't always work" objections, *no* conceptual or analytical framework is ever the answer to everything, even in a relatively small subdiscipline. This is akin to criticizing something for failing to be a thing that does not exist.

Evolutionary psychologist Daniel Dennett of Tufts University is the most cited proponent of generalized Darwinism, most notably in his 1995 book *Darwin's Dangerous Idea*, where he proposed the idea of a Darwinian process as a general algorithm involving variation, selection, and retention that is independent of context and could therefore be applied to many fields beyond biology. The book has been vigorously praised and criticized, mainly (in my opinion) because Dennett dared to invade evolutionary biology turf. The 1995 book does not mention emergence, but the basic framework is clearly an emergent one.

Darwin's name inevitably carries a lot of baggage, much of which is not Charles D.'s doing. To avoid this, I focus on highlighting the existence and importance of selection (of which, again, natural selection is but one example) in a variety of environmental systems. The emergent logic (hardly unique or original to Dennett or myself), is this: (1) variations happen; (2) some variations are more efficient, durable, stable, or otherwise favorable; and (3) the latter are more likely than others to be preserved, enhanced, and replicated.[30] Because selection is an emergent phenomenon, it is not necessary to clutter things up (as Ockham would warn us against) by claiming it to be a deterministic natural law or a goal function.

If something recurs repeatedly in nature, it is tempting to ascribe it to or think of it as a goal of nature (or of who or whatever is controlling nature). But it is not necessary to explain such features in those terms when they can be explained in simpler emergent terms, as Ockham's razor tells us we should do.

Here's an example I've used before. Does nature prefer sandstone ridgetops? Perhaps. But sandstones are more weathering resistant and physically durable than the other sedimentary rocks they often occur with. Therefore, they are preferentially preserved (selected) as denudation proceeds, and therefore plateaus and tablelands in sedimentary rocks around the world often have sandstone ridgetops. This is a simpler explanation. Here's another example. Does Mother Nature desire maximum soil moisture storage capacity in the root zone? Plant-soil interrelationships that increase soil moisture storage are beneficial to the vegetation, and therefore the plants and phenotypes that do a better job of enhancing moisture storage are privileged in terms of survival and replication. We don't need to invoke anyone's or anything's preference.

For those who do believe in God(s), Mother Earth, or some entity in charge of it all, by the way, maintaining that belief in no way requires rejection, or even skepticism, of emergent selection. If a creator does want maximum root zone soil moisture storage or sandstone ridgetops, why would they (particularly if they are wise or omniscient) not choose the simplest way to do it?

There also exists a temptation to propose that emergent selection itself is a natural law or goal function. Pascal and Pross, for instance, proposed in 2016 a "persistence principle," stated as "nature seeks persistent forms." If we simply say, "Persistent forms persist," this says the same thing without appealing to nature seeking anything.

Emergent selection is simpler than trying to ascertain and explain why sand dunes, atmospheric energy fluxes, nutrient cycles, or peat bogs, for instance, should seek or prefer anything.

THREE CASE STUDIES

Individual illustrations of the various types of selection that operate in landscapes are, I hope, clear enough. But of course, landscapes are characterized by many components linked together by multiple processes and interactions. Here are three examples of landscape evolution over different time spans and areal extents that illustrate the various types of selection going on.

Marsh Islands of the Lower Neuse River

The Neuse River rises in the Piedmont of central North Carolina and flows across the coastal plain. Its drowned lower valley is the Neuse River estuary, part of the larger Albemarle-Pamlico estuary complex. Just upstream of the town of New Bern is the fluvial-estuarine transition zone, and at the lower end of this zone there occurs marshy islands. In recent years, I have been studying the lowermost Neuse River and estuary, focusing on how it is responding to sea level rise and to major tropical cyclones, such as Hurricane Florence in 2018. For this case study, we focus on those marsh islands (Figure 6.10).

Vegetation cover on the islands is dominated by salt-tolerant herbaceous vegetation typical of brackish marshes in North Carolina, including big cordgrass (*Sporobulus cynosuroides*, formerly

Figure 6.10. Marsh islands, lower Neuse River (Google Earth™ image).

Spartina cynosuroides) and sawgrass (*Cladium jamaicense*). Some woody shrubs such as bay (*Persea* spp.) also occur, and isolated clusters of bald cypress (*Taxodium distichum*), both living and dead, occur along the island fringes.

The substrate is a dominantly organic, mucky peat overlying mineral deposits that range from sand to mud. The islands originated as sediment transported down the river was deposited as it encountered much lower slopes and backwater effects near the river mouth. This shallowing water formed sand/mud flats, some of which grew into bars and eventually islands. The uneven deposition of this sediment is a result of *gradient selection*, whereby the relatively more efficient flow paths deposited less and the more impeded paths deposited more sedimentation. Preferential flow played a role here, too, as well as resistance selection in the relative proportion of sand (bigger, heavier, deposited more readily), silt (smaller, lighter), and clay (deposited only in backwaters). But we'll focus on the bigger picture.

Once some of the deposits reached the point where they were exposed at low water, more things began living there—microbes, algae, and small nonvascular plants (*ecological filtering* at work). By providing some protective cover, and various exudates and secretions that helped bind particles together, they stabilized the features enough for larger plants, tolerant of those wet conditions, to move in. Selection processes here involved *resistance selection*, *ecosystem engineering*, and more ecological filtering. As the biological community became established, *biogeochemical selection* emerged, and plant-microbe-soil-water interactions brought *natural selection* into play via niche construction.

The relative sea level has been rising from the time the marsh islands began forming. This has gradually been pushing the locus of deposition further upstream, increasing the hydroperiod of the marshes (frequency and duration of inundation of the ground surface) and increasing salinity and making the islands more vulnerable to both marginal erosion and interior deterioration (fragmentation into vegetated wetland, mudflats, and open water). The erosion involves more resistance and gradient selection, and the changing hydrology and water chemistry selects for species better suited for those conditions. There are constant reciprocal interactions among the ecological and geomorphological dynamics, involving selection mainly in the forms of biogeochemical, ecological filtering, resistance selection, and ecosystem engineering.

Figure 6.11 shows the major stages of island development. For simplicity's sake, it is shown as a relatively simple sequence, though more complex trajectories occur. The most important forms of selection associated with the transition are shown, but other forms are at work, too. Thermodynamic selection is almost always present, at least in the background, in some form or another. Preferential flow also plays a major role at the broader scale of the evolution of the fluvial-estuarine transition zone.[31]

Another factor, the effects of fauna, is not shown in Figure 6.11. Some portions of the islands are affected by beaver (*Castor canadensis*) and nutria (*Myocaster coypus bonariensis*). Beaver is a native species that was reduced to negligible levels by trapping and hunting in the nineteenth and early twentieth centuries and has recovered in recent decades. These large rodents consume and harvest woody plants for food and to build lodges and dams, which influence flows into, through, and out of the marshes and help promote fragmentation. Nutria is a non-native introduced species (another large rodent) that consumes herbaceous marsh vegetation and digs burrows in banks, also contributing to erosion and fragmentation. In both cases, ecological filtering and ecosystem engineering selection operate.

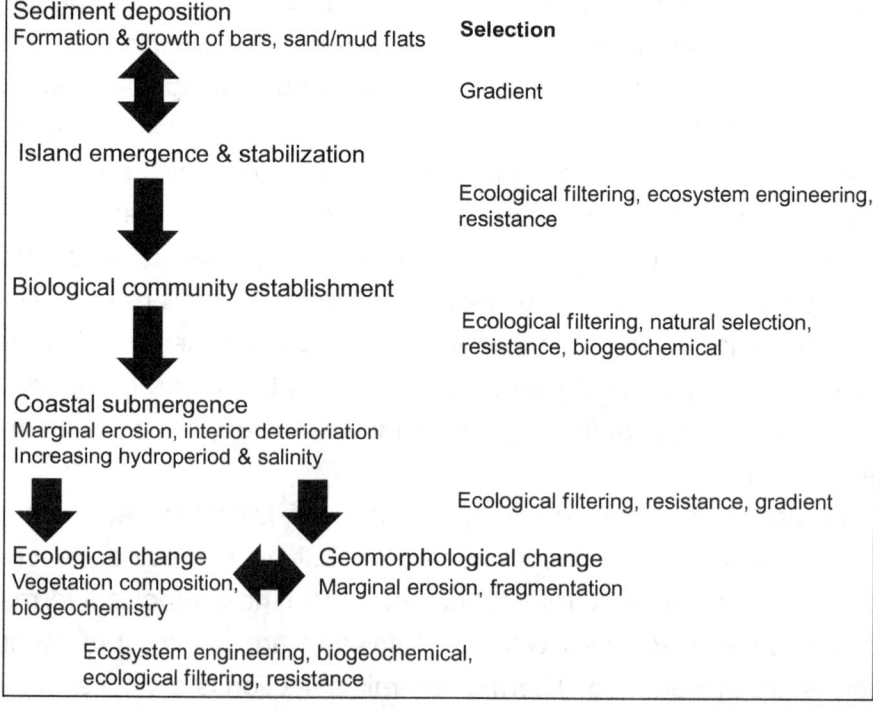

Figure 6.11. Major selection processes in the evolution of the Neuse River marsh islands.

Brazilian Cerrado

The Cerrado of central Brazil features a geographic mosaic of heterogenous vegetation growing on a very diverse array of landforms and geological settings and is the world's largest neotropical savanna (Figure 6.12). This case study is based on the work of a group of Brazilian and European scientists led by Demetrius Lira-Martins of the University of Campinos.[32] The labeling of the role of selection processes is based on my interpretation of their work, but that's it in terms of my role. Though I'd like to, I've never set foot in the Cerrado.

Cerrado refers here to a biogeographic region within Brazil that includes grasslands (regional term: *campos*), shrublands (*campos sujos*, grasslands with shrubs), savannas (grasslands with trees), and woodland savannas (closed canopy savannas, *cerradão*). The Cerrado had an original extent (much has been altered by conversion to industrial-scale agriculture) of about 2 million km^2. Lira-Martins and colleagues (who published this work in 2022) were concerned with how the complex underlying geology of the region is reflected in geomorphology and soil geography and, in turn, edaphic properties and vegetation. Their study incorporates many aspects of the interactions among geology, geomorphic processes and landforms, pedogenesis and soils, edaphic factors, vegetation cover, and biogeochemistry. Here I will focus on broad-brush landscape evolution.

As shown in Figure 6.13, denudational dissection of the plateau, guided (as always) by gradient, resistance, and network selection and preferential flow, has produced a variety of landforms, variable in elevation, slope, and drainage context. As sedimentary rocks are differentially removed, the topography exposes geological layers of different composition, resistance, and age, with the resistant ridgetops and plateau surfaces the oldest in terms of exposure to pedogenetic and ecological processes and the valley bottoms the youngest. At the local scale (Figure 6.13B), more features of

Figure 6.12. Top, middle: landscapes of the Brazilian Cerrado. Bottom: deeply weathered Oxisol in the cerrado. Photos courtesy of Diego Nascimento; photo credits, respectively, Anna Abrahãho, Rafael Silva, and Francisco Ladeira.

morphology are indicated, as well as effects of sediment transport and deposition, and of weathering (note the laterites and silcretes). These produce different sets of soil-forming factors and habitat variables, where resistance, thermodynamic, biogeochemical, and ecological selection operate. The soil/regolith and vegetation (and other biota) have a dense network of interactions, with nearly all types of selection at work.

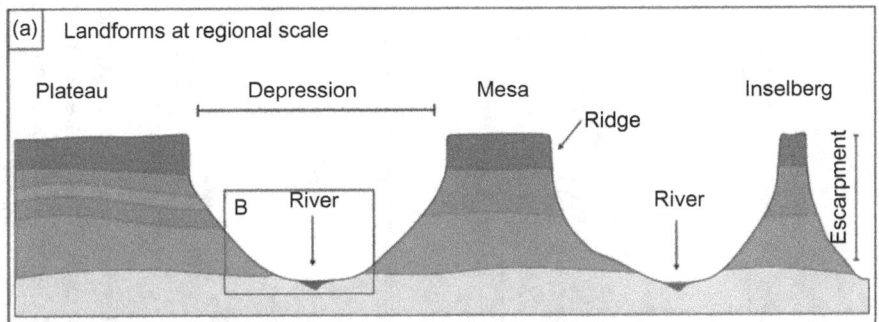

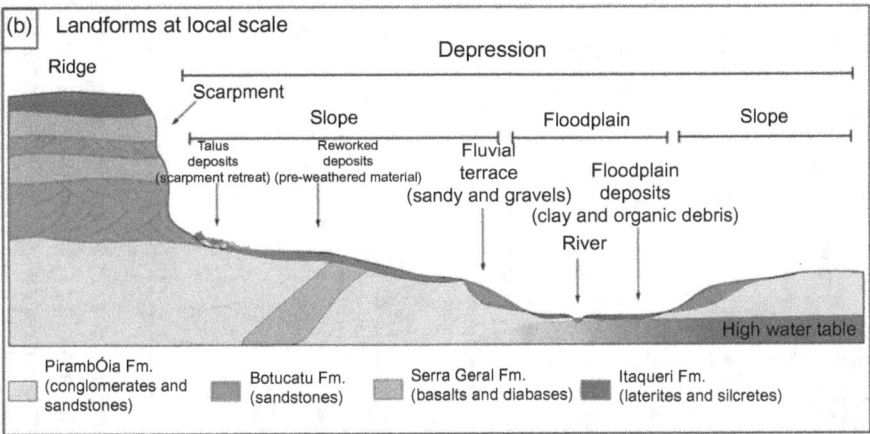

Figure 6.13. Landforms of the Cerrado. (A) Plateaus, mesas, and inselbergs covered by soil or duricrusts are common at regional scales. Landscape dissection exposes geologic materials of variable composition and resistance. (B) At local scales, the deposit of material from plateau erosion to lower portions of the slopes represents important geomorphic processes of soil formation. The material deposited next to rivers form soils in fluvial terraces and floodplains. (Reproduced with permission from D. Lira-Martins et al., "Soil Properties and Geomorphic Processes Influence Vegetation Composition, Structure, and Function in the Brazilian Cerrado," Plant and Soil 476 (2022): https://doi.org/10.1007/s11104-022-05517-y).

Figure 6.14 summarizes the main determinants of vegetation communities in the Cerrado, and Figure 6.15 indicates the major forms of selection at work.

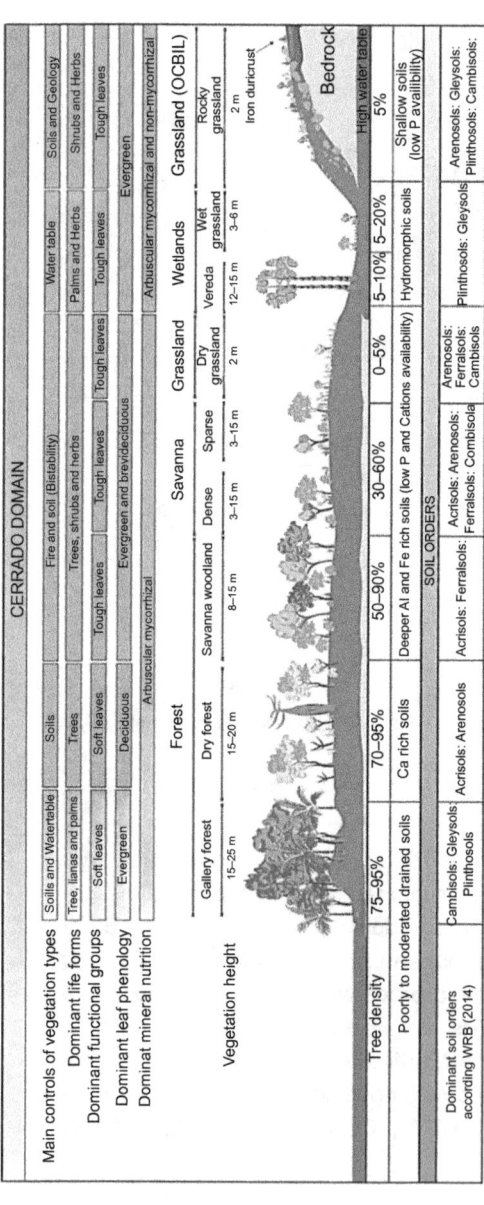
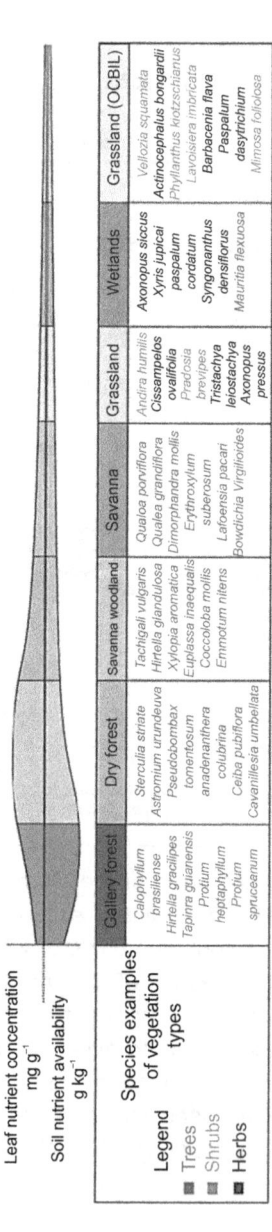

Figure 6.14. Main determinants of Cerrado vegetation types. Dominant life forms, functional groups, leaf phenology, and mineral nutrition of each vegetation type are indicated, along with characteristic soil properties and soil order of each formation are shown. Typical species are also shown. (Reproduced with permission from D. Lira-Martins et al. "Soil Properties and Geomorphic Processes Influence Vegetation Composition, Structure, and Function in the Brazilian Cerrado," Plant and Soil 476 (2022): https://doi.org/10.1007/s11104-022-05517-y, Figure 13).

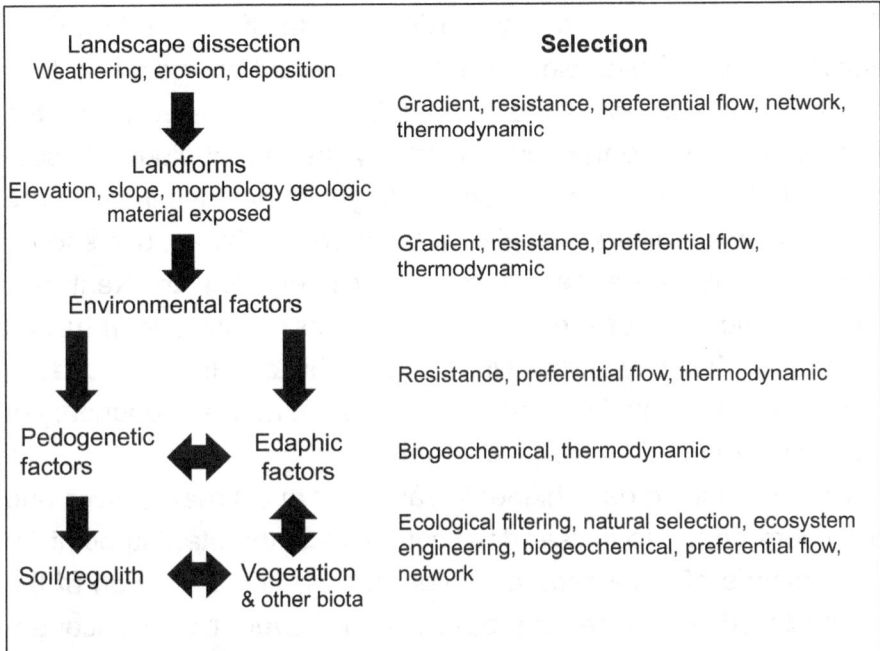

Figure 6.15. Dominant types of selection in the Brazilian Cerrado landscape. An additional ecological selection factor not shown is fire.

Inner Bluegrass, Kentucky

Like most geoscientists and ecologists, I sooner or later become fascinated by what's in my own backyard. With me it's usually sooner, and during my twenty years at the University of Kentucky I ended up studying various aspects of the fluvial, karst, and soil geomorphology, biogeomorphology, and hydrology of the Inner Bluegrass region of central Kentucky. Eventually I began putting those pieces together into a story of the coevolution of landforms, hydrological systems, soils, and ecosystems of the region, specifically focusing on the role of selection. As I've outlined this in more detail elsewhere, I'll keep it brief here.[33]

About 1.5 Ma (sometime between 1.3 and 1.8 Ma), the ancestor of the Ohio River, called the Teays, was blocked by an ice dam from a lobe of one of the great Pleistocene ice sheets. A lake formed behind the dam and eventually overflowed, diverting the

Teays to the south, to the approximate path of the modern Ohio River. The base level elevation of the diverted river did not differ much from the pre-dam Teays, but tributaries of the river from the unglaciated area to the south suddenly (geomorphologically suddenly, that is) had a shorter distance to get to their destination. The same elevation difference (give or take) divided by a much shorter distance equals a steeper slope for those rivers (e.g., the Kentucky, Cumberland, Tennessee, and Green Rivers). Rather than having to meander halfway across what is now Ohio and Indiana to reach the Teays, they could stop at what is now the northern boundary of Kentucky (marked by the Ohio River).

Steeper slope means higher shear stress and stream power, and the rivers began to incise (downcut). This is the starting point for this episode of landscape evolution, for the fluvial incision of the Kentucky River lowered the base level for every hydrological entity in the Inner Bluegrass and thereby helped drive evolution of landforms, soils, and ecosystems, assisted by climate shifts and windblown dust. As the river sawed through the limestones, tributaries began eroding downward, too. The lowering base level also stimulated karst development—vertical and horizontal joints were enlarged, resulting in development on the surface of swallets, dolines, karst valleys, and other features, and of caves and conduits belowground (Figure 6.16).

Different parts of the landscape became dominated by either karst or fluvial processes, depending on their setting relative to the incising streams and sometimes on happenstance—since, once initiated, either karst or fluvial pathways tend to persist. For a while, that is; transitions back and forth between karst and fluvial features occurred. Traditional perceptions of fluviokarst landscape evolution were founded on succession-like patterns ending up in a transition from karst to fluvial domination (or vice versa). However, in central Kentucky the competition between karst and fluvial, or subsurface and surface, processes does not necessarily result in steadily increasing domination of one or the other. Instead, karst-to-fluvial and fluvial-to-karst transitions are common at the local

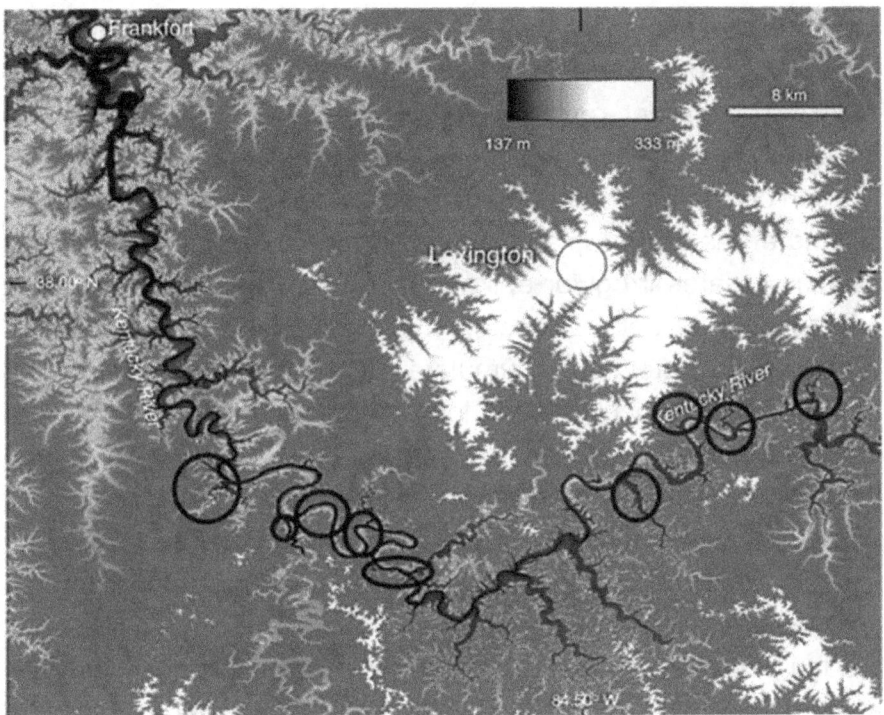

Figure 6.16. Kentucky River gorge area of the Inner Bluegrass. Circles and ovals highlight main sites where fieldwork examining landscape evolution driven by Kentucky River incision was conducted.

scale, and the same landscape often contains evidence of both kinds of transitions.

Except for river alluvium, all soils in the area have weathered limestone parent material. These are either relatively pure phosphatic or non-phosphatic limestone interbedded with shale and siltstones. Glacially derived loess (windblown silt) deposited around 25–12 ka coats some of the area. Dust deposition was likely nonuniform, because everything is, but differences within the landscape are primarily related to subsequent redistribution by erosion and deposition. Silty cover beds range from nonexistent to >1 m thick.

Differences in soils are primarily related to parent material, topographically controlled variations in erosion status, presence of silt cap (or soil surface formed from loess deposits), and mollic properties. Mollic soil properties are associated with organic-rich surface horizons with high cation exchange capacity. The chemically basic parent material makes the latter criterion easy to achieve, so where organic matter can accumulate, mollic properties can occur. In the contemporary landscape, this is most often on sloping areas of the uplands or valley sides. In many cases, dissolution creates irregular epikarst layers just below the soil, with related surface depressions (sometimes just centimeters deep) where organic litter can accumulate. On the flatter upper surfaces, conversion to agriculture has long since occurred, and some of the pre-European vegetation was grassland, savanna, or woodland communities. On the steeper slopes, however, forest cover was often maintained, allowing more organic matter accumulation.

Figure 6.17 shows landscape relationships for some common soils. The silt caps are present mainly in the uneroded or barely eroded soils, and the loess has elsewhere been largely washed away. The mollic soils are found chiefly on the side shoulders and upland edges, and erosion status is predictably related to topography.

Plant-soil-parent material interactions, preferential flow, and resistance selection are particularly important in karst landscapes, and especially with respect to differential weathering, apportioning of hydrological flows, and the role of roots in exploiting rock joints and fractures and widening them, via funneling of water, formation of organic acids, root respiration of CO_2, and microbial activity in the rhizosphere.

Inner Bluegrass Selection

How does selection operate to produce the Inner Bluegrass landscape? Gradient selection and resistance selection influence both fluvial and karst processes, particularly the river incision that drives fluvial-to-karst and karst-to-fluvial transitions (and keeps both sets of processes active). This constant rejuvenation preserves spatial

Pure limestone: Maury, McAfee   Silt cap: Maury
Interbedded limestone: Faywood, Lowell, Fairmount
Alluvium: Huntingdon

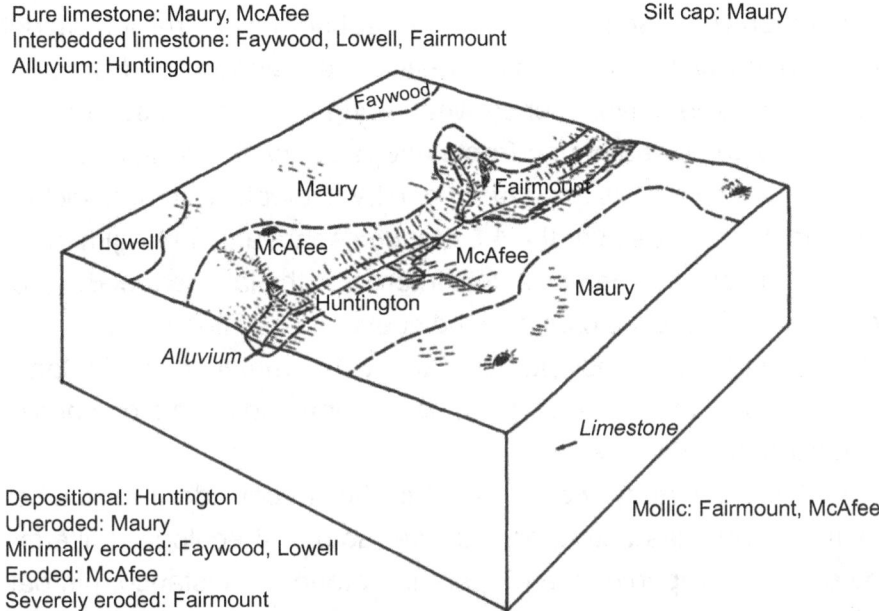

Depositional: Huntington
Uneroded: Maury   Mollic: Fairmount, McAfee
Minimally eroded: Faywood, Lowell
Eroded: McAfee
Severely eroded: Fairmount

Figure 6.17. Landscape block diagram showing how soil types relate to topography, parent material, erosion, loess (silt) deposits, and mollic properties. Base block diagram is reproduced from the Soil Survey of Jessamine and Woodford Counties (https://casoilresource.lawr.ucdavis.edu/sde/ ?series=faywood#block-diagrams).

structuring processes and inhibits development of very old, mature soils, which cannot form without long, uninterrupted progressive pedogenesis.

Though groundwater flows are driven by gravity, they are strongly directed by rock structures such as joints, bedding planes, fractures, and fissures. The larger, steeper, and smoother of these are more likely to convey flow. Positive feedback reinforces these pathways by dissolutional enlargement, just as erosional enlargement of channels reinforces those surface flow paths.

Selection and spatial differentiation lead to distinctive soil properties as pedogenesis proceeds in the region. Examples include mollic properties, sufficient base saturation to qualify as alfisols or mollisols (as opposed to the more acidic soils that usually form on non-carbonate parent material in the larger region), accumulation of iron and aluminum oxides, and formation of iron and manganese concretions (and sometimes compact subsoil layers that inhibit root penetration called fragipans). Resistance selection and gradient selection powerfully influence weathering in the epikarst, creating highly irregular weathering fronts and regolith/rock contacts.

Biotic communities are determined by ecological filtering, including climate change over the 1.5 Ma time span, and fire regimes, as well as the environmental factors internal to the landscape. Biogeochemical selection is crucial in the coevolution of floral, faunal, and microbial communities and soil. Table 6.1 summarizes major landscape evolution processes and phenomena and the major types of selection affecting them.

In this case study, we'll go a bit further to consider multiple selection processes acting on multiple factors. If ecological filtering by edaphic properties were operating alone, for instance, pin oak (*Quercus palustris*), an iconic tree in Lexington, would be rare in the region (though it is native). Pin oak has low calcium carbonate tolerance and fares poorly with soil pH above 6.5. This makes it poorly suited to the region's soils. But pin oak is common in urban and suburban sites on account of resistance selection and positive artificial selection. Pin oak tolerates urban settings in general and air pollutants such as sulfur dioxide—an important consideration in Kentucky's (unfortunately and overwhelmingly) coal-fired economy. The tree is widely planted for landscaping of parks, lawns, and streets.

Another example is the strong influence of joints and faults on topography via resistance and gradient selection. But if these were the only significant selection processes, landforms and topography would mirror the underlying structures. Sometimes they do, as for example when dolines (sinkholes) are lined up along a fault structure. But such mirroring is not always evident, and dolines (for instance) are scattered across the landscape as well as along fault traces.

Figure 6.18 shows how selection of various aspects of the landscape may be overlapping or mutually reinforcing in some cases and offsetting or mutually interfering in others. For example, the selection processes influencing slope erosion counteract those

Table 6.1. KEY ASPECTS OF LANDSCAPE EVOLUTION IN THE INNER BLUEGRASS REGION OVER THE PAST 1.5 MA AND MAJOR TYPES OF SELECTION APPLICABLE

Process	Description	Selection	Outcome
Fluvial dissection	Kentucky River downcutting and tributary incision	Gradient, resistance, preferential flow	Development and growth of channel network and watersheds; lowering of base levels
Karstification	Dissolutional denudation	Gradient, resistance, preferential flow	Development of conduits, caves, and subsurface flow networks; formation of major karst surface forms (e.g. dolines)
Slope erosion	Soil erosion and mass wasting on uplands and valley side slopes	Gradient, resistance, preferential flow	Redistribution of soil and sediment; variable soil depth
Loess redistribution	Net removal or accumulation of Pleistocene loess deposits	Gradient, resistance, preferential flow	Modification of soil morphology and chemistry
Local weathering	Rock weathering within or beneath regolith cover	Resistance, preferential flow	Formation of karst microtopography; variable thickness of regolith; irregular weathering front geometry

Continued

Table 6.1. Continued

Process	Description	Selection	Outcome
Runoff	Local partitioning of precipitation into infiltration, surface and subsurface runoff	Gradient, resistance, preferential flow	Transitions between karst/subsurface and fluvial/surface landforms and hydrology
Vegetation	Establishment of vegetation communities	Ecological filtering, biogeochemical, resistance	Development of vegetation cover and faunal communities; emergence of characteristic plant, soil and landform interactions and relationships
Root zone	Development of root accommodation space, moisture holding capacity, and rhizosphere in soil, regolith, and rock	Biogeochemical, resistance, preferential flow	Average, net increase of soil depth and moisture storage; initiation and enhancement of biogeochemical cycles
Pedogenesis	Soil formation; emergence of soil profiles and properties	Resistance, preferential flow, ecological filtering	Development of soil cover and soil, landform, and biotic relationships

affecting progressive pedogenesis, whereas selection in local weathering enhances pedogenesis. Selection in root zone development, however, is mutually reinforcing with the selection at work in vegetation and pedogenesis.

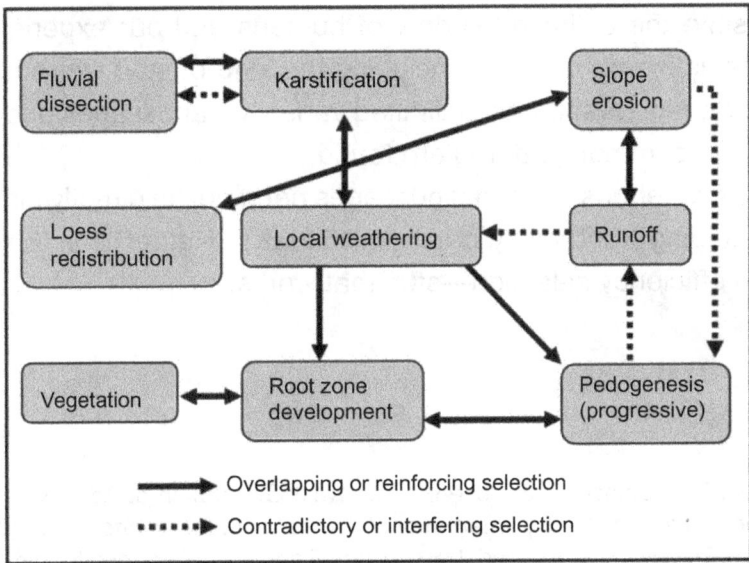

Figure 6.18. Interactions among key processes in the Inner Bluegrass landscape reflecting reinforcing or overlapping selection and contradictory or opposing selection effects. Other interactions where selection is not a major factor are not shown.

SUMMARY: STUFF THAT WORKS

Legendary songwriter Guy Clark (1941–2016) never explicitly addressed landscape evolution or selection, but he certainly provided insight into the mysterious ways of human experience. With Rodney Crowell, he wrote a song called "Stuff That Works." Here's the chorus:

> Stuff that works, stuff that holds up
> The kind of stuff you don't hang on the wall
> Stuff that's real, stuff you feel
> The kind of stuff you reach for when you fall

Things (including relationships and connections) that are well made, useful, and durable tend to last. Things that get the job done, and do so efficiently, are preferred and selected by humans or non-human nature. Stuff that works is invented or arises independently all over the world.

Despite the unlimited variety of humans and our experiences, there exist commonalities, including the kind of stuff we reach for when we fall. Despite the unlimited variety of landscapes and their histories, common patterns are found.

The mysterious ways of landscapes need not be directly dictated by a guiding hand or by deterministic laws. Rather, they emerge due to efficiency selection—stuff that works.

NOTES

1. See, for instance: J. J. Schenk, "The Next Generation of Adaptive Radiation Studies in Plants," *International Journal of Plant Sciences* 182 (2021): 245–262; J. T. Stroud and J. B. Losos, "Ecological Opportunity and Adaptive Radiation," *Annual Review of Ecology*, Evolution, and Systematics 47 (2016): 507–532; G. A. Wellborn and R. B. Langerhans, "Ecological Opportunity and the Adaptive Diversification of Lineages," *Ecology and Evolution* 5 (2015): 176–195.

2. D. H. Erwin and S. Tweedt, "Ecological Drivers of the Ediacaran-Cambrian Diversification of Metazoa," *Evolutionary Ecology* 26 (2012): 417–433.

3. G. J. Retallack, "Woodland Hypothesis for Devonian Tetrapod Evolution," *Journal of Geology* 119 (2011): 235–258.

4. R. C. van de Visjel, J. van Belzen, and T. J. Bouma, "Estuarine Biofilm Patterns: Modern Analogues for Precambrian Self-Organization," *Earth Surface Processes and Landforms* 45 (2020): 1141–1154.

5. For example: F. P. D. Cotterill and M. J. De Wit, "Geoecodynamics and the Kalahari Epeirogeny: Linking Its Genomic Record, Tree of Life and Palimpsest into a Unified Narrative of Landscape Evolution," *South African Journal of Geology* 114 (2011): 489–514; V. Hoffman, G. A. Verboom, and F. P. D. Cotterrill, "Dated Plant Phylogenies Resolve Neogene Climate and Landscape Evolution in the Cape Floristic Region," *PLoS One* 10 (2014): https://doi.org/10.1371/journal.pone.0137847.

6. J. W. Crawford et al., "Microbial Diversity Affects Self-Organization of the Soil—Microbe System with Consequences for Function," *Journal of the Royal Society Interface* 9 (2012): 1302–1310; E. Laliberté et al., "How Does Pedogenesis Drive Plant Diversity?" *Trends in Ecology and Evolution* 28 (2013): 331–340; C. C. Pregitzer, J. K. Bailey, S. C. Hart, and

J. A. Schweitzer, "Soils as Agents of Selection: Feedbacks between Plants and Soils Alter Seedling Survival and Performance," *Evolutionary Ecology* 24 (2010): 1045–1059; J. A. Schweitzer et al., "Are There Evolutionary Consequences of Plant-Soil Feedbacks along Soil Gradients?" *Functional Ecology* 28 (2014): 55–64; I. M. Ware et al., "Feedbacks Link Ecosystem Ecology and Evolution across Spatial and Temporal Scales: Empirical Evidence and Future Directions," *Functional Ecology* 33 (2019) 31–42.

7. P. S. Eagleson, *Ecohydrology: Darwinian Expression of Vegetation Form and Function* (New York: Cambridge University Press, 2021); A. G. Lapenis, "Directed Evolution of the Biosphere: Biogeochemical Selection or Gaia?" *Professional Geographer* 54 (2022) 379–391; J. D. Phillips, "Goal Functions in Ecosystem and Biosphere Evolution," *Progress in Physical Geography* 32 (2008): 51–64.

8. H. Q. Huang and G. C. Nanson, "Hydraulic Geometry and Maximum Flow Efficiency as Products of the Principle of Least Action," *Earth Surface Processes and Landforms* 25 (2000): 1–16; H. Q. Huang and G. C. Nanson, "Why Some Alluvial Rivers Develop an Anabranching Pattern," *Water Resources Research* 43 (2007): W07441; G. C. Nanson and H. Q. Huang "Least Action Principle, Equilibrium States, Iterative Adjustment and the Stability of Alluvial Channels," *Earth Surface Processes and Landforms* 33 (2008): 923–942; G. C. Nanson and H. Q. Huang "Self-Adjustment in Rivers: Evidence for Least Action as the Primary Control of Alluvial-Channel Form and Process," *Earth Surface Processes and Landforms* 42 (2017): 575–594.

9. G. C. Nanson and H. Q. Huang "A Philosophy of Rivers: Equilibrium States, Channel Evolution, Teleomatic Change and the Least Action Principle," *Geomorphology* 302 (2018): 3–19.

10. R. Pascal and A. Pross, "The Logic of Life," *Origins of Life and Evolution of the Biosphere* 46 (2016): 507–513; A. Pross and R. Pascal, "How and Why Kinetics, Thermodynamics, and Chemistry Induce the Logic of Biological Evolution," *Beilstein Journal of Organic Chemistry* 13 (2017): 665–674.

11. R. C. Dewar, "Maximum Entropy Production and Non-Equilibrium Statistical Mechanics," in *Non-Equilibrium Thermodynamics and the Production of Entropy. Understanding Complex Systems*, edited by A. Kleidon and R. D. Lorenz (Berlin: Springer, 2005), 41–55; E. N. Lorenz, "Generation of Available Potential Energy and the Intensity of the General Circulation," in *Dynamics of Climate*, edited by R. F. Pfeffer (Tarrytown, NY: Permagon, 1960), 86–92; H. Ozawa, A. Ohmura, R. D. Lorenz, and T. Pujol, "The Second Law of Thermodynamics and the Global Climate System: A Review of the Maximum Entropy Production Principle," *Reviews of Geophyics*. 41 (2003): 1018; G. W. Paltridge, "Global Dynamics and Climate: As a System of Minimum Entropy Exchange," *Quarterty Journal of the Royal Meteorology Society* 101 (1975): 475–484.

12. A. Bejan, "Constructal Theory of Pattern Formation," *Hydrology and Earth System Sciences* 11 (2007): 753–768; A. Bejan, "Evolution in Thermodynamics," *Applied Physics. Review* 4 (2017): https://doi.org/10.1063/1.4978611.

13. Examples include: A. Brebner and K. C. Wilson, "Derivation of the Regime Equations from Relationships for Pressurized Flow by use of the Principle of Minimum Energy-Degradation Rate," *Proceedings, Institute of Civil Engineers* 36 (1967) 47–62; T. R. H. Davies and A. J. Sutherland, "Resistance to Flow Past Deformable Boundaries," *Earth Surface Processes and Landforms* 5 (1980): 175–179; A. Kleidon, Y. Malhi, and P. M. Cox, "Maximum Entropy Production in Environmental and Ecological Systems," *Philosophical Transactions of the Royal Society* B 365 (2010), 1297–1302; P. Molnar and J. A. Ramirez, "Energy Dissipation Theories and Optimal Characteristics of River Networks," *Water Resources Research* 34 (1998): 1809–1818; K. Paik and P. Kumar, "Optimality Approaches to Describe Characteristic Fluvial Patterns on Landscapes," *Philosophical Transactions of the Royal Society* B 365 (2010): 1387–139; T. R. Smith, "A Theory for the Emergence of Channelized Drainage," *Journal of Geophysical Research—Earth Surface* 115 (2010): https://doi. org/10.1029/2008JF001114; C. T. Yang, "Potential Energy and Stream Morphology," *Water Resources Research* 7 (1971): 311–322; C. T. Yang, C. C. S. Song, and M. J. Woldenberg, "Hydraulic Geometry and Minimum Rate of Energy Dissipation," *Water Resources Research* 17 (1981): 1014–1018.

14. Y. Liu, T. S. Steenhuis, and J.-Y. Parlange, "Formation and Persistence of Fingered Flow Fields in Coarse-Grained Soils under Different Moisture Contents," *Journal of Hydrology* 159 (1994): 187–195; C. J. Ritsema, L. W. Dekker, J. L. Nieber, and T. S. Steenhuis, "Modeling and Field Evidence of Finger Formation and Finger Recurrence in a Water Repellent Sandy Soil," *Water Resources Research* 34 (1998): 555–567; C. J. Ritsema et al., "Recurring Fingered Flow Pathways in a Water Repellent Sandy Soil," *Hydrology and Earth System Sciences* 4 (1997): 777–786.

15. For instance, see: M. Filipponi, P.-Y. Jeannin, and L. Tacher, "Evidence of Inception Horizons in Karst Conduit Networks," *Geomorphology* 106 (2009): 86–99; S. R. H. Worthington, "How Preferential Flow Delivers Pre-Event Groundwater Rapid to Streams," *Hydrological Processes* 33 (2019): 2373–2380.

16. Reviews and syntheses of preferential flow in soils are given by Hendrickx and Flury (2001), Gerke (2006), and Guo and Lin (2018). As usual, scale is important, and different nonuniform flow processes are critical at different scales (Hendrickx and Flury, 2001). For example, vertical preferential flow was found to be more important at the plot scale in Glaser et al.'s (2019) simulations, while at the catchment scale, rapid lateral subsurface flow was more important. References: H. H. Gerke, "Review Article: Preferential Flow Descriptions for Structured Soils," *Journal of Soil Science and Plant Nutrition* 169 (2006): 382–400; B. Glaser, C. Jackisch, L. Hopp, and J. Klaus, "How Meaningful Are Plot-Scale Observations of Preferential Flow for Catchment Models?" *Vadose Zone Journal* 18 (2019): 180,146; L. Guo and H. Lin, "Addressing Two Bottlenecks to Advance the Understanding of Preferential Flow in Soils," *Advances in Agronomy* 147 (2018): 61–117;

J. M. H. Hendrickx and M. Flury, "Uniform and Preferential Flow Mechanisms in the Vadose Zone," in *Conceptual Models of Flow and Transport in Fractured Rock*, ed. National Research Council (Washington, DC: The National Academies Press, 2001), 149–187.

17.   J. D. Phillips et al., "Deterministic Uncertainty and Complex Pedogenesis in Some Pleistocene Dune Soils. *Geoderma* 73 (1996): 147–164. My co-authors here were students in my physical geography graduate seminar at East Carolina University.

18.   S. Franklin, B. Vasilas, and Y. Jin, "More Than Meets the Dye: Evaluating Preferential Flow Paths as Microbial Hotspots," *Vadose Zone Journal* 18 (2019): 190,024; B. Li et al., "Preferential Flow in the Vadose Zone and Interface Dynamics: Impact of Microbial Exudates," *Journal of Hydrology* 558 (2018): 72–89.

19.   W. F. Doolittle, "Natural Selection through Survival Alone, and the Possibility of Gaia," *Biology and Philosophy* 29 (2014): 415–423; W. F. Doolittle, "Darwinizing Gaia," *Journal of Theoretical Biology* 434 (2017): 11–19; T. M. Lenton et al., "Selection for Gaia across Multiple Scales," *Trends in Ecology and Evolution* 33 (2018): 633–645; J. Toman and J. Flegr, "Stability-Based Sorting: The Forgotten Process Behind (Not Only) Biological Evolution," *Journal of Theoretical Biology* 435 (2017): 29–41.

20.   M. J. Kirkby, "Hillslope Process-Response Models Based on the Continuity Equation," *Institute of British Geographers Special Publication* 3 (1971): 15–30; I. Rodriguez-Iturbe and A. Rinaldo, *Fractal River Basins. Chance and Self-Organization* (New York: Cambridge University Press, 1997); M. J. Woldenberg, "Spatial Order in Fluvial Systems: Horton's Laws Derived from Mixed Hexagonal Hierarchies of Drainage Basin Areas," *Geological Society of America Bulletin* 80 (1969): 97–112. Kirkby and Woldenberg are both among my early academic heroes, and I had a chance to meet both several times over the years. Mike Woldenberg was very supportive and encouraging to me in my work at a time when I was having trouble getting people to take me seriously.

21.   A. J. Lotka, "Contributions to the Energetics of Evolution," *Proceedings, National Academy of Sciences (USA)* (1922): 147–151.

22.   B. D. Fath, S. E. Jorgensen, B. C. Patten, and M. Straskraba, "Ecosystem Growth and Development," *BioSystems* 77 (2004): 213–228; H. Lin, "Thermodynamic Entropy Fluxes Reflect Ecosystem Characteristics and Succession," *Ecological Modelling* 298 (2015): 75–86.

23.   Eagleson: see note 7. E. Lekevicius, "The Russian Paradigm in Ecology and Evolutionary Biology: Pro et Contra," *Acta Zoologica Litania* 16 (2006): 3–19; W. H. Verboom and J. S. Pate, "Exploring the Biological Dimension to Pedogenesis with Emphasis on the Ecosystems, Soils and Landscapes of Southwestern Australia," Geoderma 211/212 (2013): 154–183.; D. M. Wilkinson, "The Fundamental Processes in Ecology: A Thought Experiment on Extraterrestrial Biospheres," *Biological Reviews* 78 (2003): 171–179.

24. An alternative perspective is given by Allen Hunt and Stefano Manzoni, who argue that that although the plant roots cannot know what the globally optimal paths are, if they set up a potential difference over the soil, the water flow to the roots will follow those optimal paths, and if such paths intersect nutrients, the nutrient signal can draw the plant roots up the flow. See *Networks on Networks: The Physics of Geobiology and Geochemistry* (IOPP Publishing, 2016). Hunt has published extensively on this general topic; one example is A. G. Hunt, "Spatio-Temporal Scaling of Vegetation Growth and Soil Formation from Percolation Theory," *Vadose Zone Journal* 15 (2016): https://doi.org/10.2136/vzj2015.01.0013.

25. T. Jerin and J. D. Phillips, "Local Efficiency in Fluvial Systems: Lessons from Icicle Bend," *Geomorphology* 282 (2017): 119–130. Tasnuba Jerin was a PhD student I worked with at the University of Kentucky.

26. S. F. Gallen, "Lithologic Controls on Landscape Dynamics and Aquatic Species Evolution on Post-Orgenic Mountains," *Earth and Planetary Science Letters* 493 (2018): 150–160.

27. D. Corenblit et al., "Control of Sediment Dynamics by Vegetation as a Key Function Driving Biogeomorphic Succession within Fluvial Corridors," *Earth Surface Processes and Landforms* 34 (2009): 1790–1810; D. Corenblit, J. Steiger, A. M. Gurnell, and E. Tabacchi, "Darwinian Origin of Landforms," *Earth Surface Processes and Landforms* 32 (2007a): 2070–2073; D. Corenblit, E. Tabacchi, J. Steiger, and A. M. Gurnell, "Reciprocal Interactions and Adjustments between Fluvial Landforms and Vegetation Dynamics in River Corridors: A Review of Complementary Approaches," *Earth Science Reviews* 84 (2007b): 56–86; J. Eichel, D. Corenblit, and R. Dikau, "Conditions for Feedbacks between Geomorphic and Vegetation Dynamics on lateral moraine Slopes: A Biogeomorphic Feedback Window," *Earth Surface Processes and Landforms* 41 (2016): 406–419.

28. A. Larsen, J.-H May, and X. Carah, "Late Quaternary Biotic and Abiotic Controls on Long-Term Sediment Flux in a Northern Australian Tropical River System," *Earth Surface Processes and Landforms* 44 (2019): 2494–2509.

29. A. V. de Groot et al., "Tales of Island Tails: Biogeomorphic Development and Management of Barrier Islands," *Journal of Coastal Conservation* 21 (2017): 409–419; J. Eichel, M. S. Krautblatter, and R. Dikau, "Biogeomorphic Interactions in the Turtmann Glacier Forefield, Switzerland," *Geomorphology* 201 (2013): 98–110; S. N. Lane, L. Borgeaud, and P. Vittoz, "Emergent Geomorphic-Vegetation Interactions on a Subalpine Alluvial Fan," *Earth Surface Processes and Landforms* 41 (2016): 72–86; J. D. Phillips et al., "Domination of Hillslope Denudation by Tree Uprooting in an Old-Growth Forest," Geomorphology 276 (2017): 27–36.

30. A similar framework was outlined by R. Khatibi in his effort to integrate systems theory and evolutionary thinking in the context of flood risk. He pointed out that some system configurations may prime a new entity with emergent properties that confer a selective advantage. Once selected, such properties and configurations typically survive via positive feedback. Consistent

with the probabilistic nature of selection, Khatibi sees selection as a potential, not an imperative. R. Khatibi, "Evolutionary Systemic Modeling of Practices on Flood Risk," *Journal of Hydrology* 401 (2011): 36–52.

31. J. D. Phillips, "Geomorphology of the Fluvial-Estuarine Transition Zone, Neuse River, North Carolina," *Earth Surface Processes and Landforms* 47 (2022): 2044–2061; J. D. Phillips, "Store and Pour: The Evolution of Flow Systems in Landscapes," *Catena* 216 (2022): 106,357.

32. D. Lira-Martins et al., "Soil Properties and Geomorphic Processes Influence Vegetation Composition, Structure, and Function in the Brazilian Cerrado," *Plant and Soil* 476 (2022): https://doi.org/10.1007/s11104-022-05517-y.

33. This story and the supporting references are given in more detail in Chapter 8 of *Landscape Evolution* and in J. D. Phillips, "Place Formation and Axioms for Reading the Natural Landscape," *Progress in Physical Geography* 42 (2018): 697–720.

7

Store and Pour and Adapt

The mysterious ways of nature are such that despite an endless variety of different combinations of environmental controls and influences, and notwithstanding an uncountable number of potential evolutionary trajectories, certain patterns and structures recur again and again. Remarkably, these recurring features are not dictated by any specific laws of nature or universal principles. Rather, they are emergent products of efficiency selection. Perhaps no natural phenomenon illustrates this better than the evolution of hydrological flow systems.

When more precipitation falls than plants can use and the ground can hold, it must go somewhere. That inevitable redistribution is driven by gravity. We've already covered how various forms of efficiency selection promote preferential flow of various kinds. Here we'll move on to how and why hydrological systems—dare I say it—adapt to meet geophysical and ecological needs.

Hydrological systems are inextricably intertwined with ecological, pedological, and geomorphological ones. Movements and transformations of water perform/reflect transfers of mass and energy, the work of biological metabolism, weathering, and transport of material, and biogeochemical cycling. Thus, the evolution of the hydrological components of landscapes is a good reflection as of how landscapes in general develop and function.

Mysterious Ways. Jonathan D. Phillips, Oxford University Press. © Oxford University Press (2025).
DOI: 10.1093/9780197755129.003.0007

FILLING AND SPILLING

Filling and spilling are easiest to see in streams that occasionally run dry or very low. The pools in the stream bed become disconnected. When runoff increases, the pools gradually fill, until they spill downstream, and the pools are reconnected. The same phenomenon happens in many other situations (Figure 7.1). For instance, some of the early work on preferential soil moisture flow found that the preferential flow paths (PFPs) often form when pockets of subsurface storage (e.g., pores or voids, layers, or pockets of organic matter) become saturated (fill) and begin exporting flow (spill).

Figure 7.1. Simple examples of fill and spill. Pools in this South Australia stream are isolated during dry periods. When they fill and overflow, there is continuous flow within the channel. On the right, on this Czech hillslope when this slight depression becomes saturated, water begins spilling downslope as seepage.

Filling of storage elements so that they begin transferring water is a common and critical phenomenon with respect to crossing thresholds for runoff generation, multichannel flow or between different flow regimes, and for establishing and expanding hydrological connectivity. Fill-and-spill terminology has been used in the hydrology literature since at least the early 1990s. Fill-and-spill dynamics are often an important part of store and pour, but the latter is a broader

concept that also applies at dry times when there is no filling going on and to wet periods when all normal storage components are
full. Fill and spill is primarily concerned with temporal dynamics,
while store and pour (S&P) is primarily concerned with the structure of hydrological systems with respect to their capacities to store
moisture and transport or export it via both slower and more rapid
pathways.

As we shall see, however, filling and spilling is an important
process in the transitions in flow and storage dynamics that comprise S&P.

EXAMPLES OF STORE-AND-POUR MORPHOLOGY

S&P is a shorthand for the tendency of hydrological systems to
organize themselves into two broad elements. One is a "pour" component comprised of channels, conduits, connected pores, and
other PFPs—the most efficient avenues for moving water. These
are the portions of the system that most often convey flow (and
at the most rapid rates). The "store" components are where water
is retained—sometimes remaining in place, but if moving, always
more slowly than in the pour elements. The soil matrix, porous rock
and small fractures in groundwater systems, and inter-channel areas of surface watersheds are all examples. Storage zones also
function as sources of water to flow paths. They export water to
PFPs when the storage capacity is filled (fill and spill) or when inputs exceed their intake capacity—for instance, when it is raining
faster than the water can soak in (infiltrate), a phenomenon called
infiltration-excess or Hortonian runoff.

I began thinking about S&P as a product of efficiency selection, and as one of the mysterious ways that nature produces
similar phenomena in very different situations, due to a combination of out-in-the-boonies field observations and the published
scientific literature. I had been spending a great deal of both recre-
ational and professional time (these are not always distinct in my

world) in swamps of the fluvial-estuarine transition zones of lower coastal plain rivers of the Carolinas. In particular, I was interested in how and why these typically multichannel systems formed and the hydrological roles of various channels. The formation of multi-channel rather than single-channel patterns typically occurs where there are high sediment loads relative to river transport capacity. In the lower coastal plain, this often occurs due to low transport capacity because of the low slopes, but these systems are also found where sediment loads are extremely low (so-called blackwa-ter streams characterized by clear, dark water stained by organic acids) and where floodplain sediments are dominantly organic (with limited indications of much mineral sediment input). I had also seen that in these channel-wetland complexes many—in fact, most—channels had multiple roles in terms of hydrological flux. You have your main channels, which usually convey flow down-stream and always have water, but where flow sometimes comes to a near stop or reverses on incoming tides or during storm surges. You have your secondary channels that are usually backwater-flooded from the main channels but can pour flow downstream during high-water periods. Other channels are disconnected lakes (sloughs) much of the time but become connected and convey wa-ter at high stages. And there were channels that normally have little or no surface water and no flow that become activated in floods (Figure 7.2).

While this swamp business was going on, I read Stephen Wor-thington's article about the evolution of dual porosity (a name applied to may S&P configurations) in karst environments and its role in solute fluxes and runoff responses. It was no great reach to connect it with the body of work on preferential flow and dual porosity in soils. This in turn connected to the idea that these are conceptually (and functionally) similar to surface drainage networks consisting of channels and interfluves.[1] It occurred to me that the lower coastal plain fluvial-estuarine transition zone channel-wetland complexes could be doing essentially the same thing.

Figure 7.2. Environments of the channel-wetland complex of the lower Neuse River, North Carolina. Clockwise from upper left: main river channel; backwater subchannel conveying flow and hydraulically connected with adjacent swamps during high flow; swamp adjacent to main river channel; upper reaches of a backwater channel that serves as a spillway at higher flows; flow exchange between active channel (to right) and normally isolated slough, with beaver lodge; depressional floodplain lake connected to channels and conveying flood during floods.

Several studies of PFPs in soil noted that the development of S&P configurations is highly advantageous to plants and other biota, providing a means to slow or retain moisture for use by vegetation but also mechanisms for moving the excess. Further, plant-soil interactions tend to facilitate this trend, indicating a likely role for ecosystem engineering and selection.[2] This in turn called to mind broader theories of the development of flow systems, such as constructal theory and percolation-theory based ideas.[3] These link together subsurface and surface flow networks (and other types of flux networks) and sent me back to more familiar territory of stream networks. Do surface watersheds, hillslopes, soil moisture, groundwater systems, and channel-wetland complexes share some common "mysterious way" that causes them to develop S&P patterns?

As it turns out, yes.

STREAM NETWORKS AND DRAINAGE BASINS

We've already discussed how gradient, resistance, and network selection often produce branching, dendritic forms. The similar topographic and topological architectures in many different climates, geologic settings, and biogeographic settings are remarkable given the different environmental contexts. Those similarities suggest some unifying underlying principles or laws that are independent of local and regional environmental controls, leading to studies of this phenomenon not only by hydrologists, geomorphologists, geographers, and geologists but also by mathematicians, systems theorists, physicists, and philosophers. A key outcome, as discussed in Chapter 6, was the discovery that a branching, dendritic network is a maximum-efficiency configuration for any system of gathering fluids from an area and delivering it to a central location.

Beneath the ground surface, probably every hydrologist and soil scientist (not to mention farmers, miners, excavator operators, and ditch diggers) recognized that Darcian flow-through-porous-media is not necessarily the norm, and that preferential flow is common. Nevertheless, Darcian frameworks and associated models have traditionally dominated groundwater and soil hydrology. Though it was recognized that Darcian flow conditions are often a major oversimplification, it was frequently assumed, sometimes with justification, that a flow-through-porous-media representation averages out variations associated with nonuniform flow well enough for many purposes. That has changed in the past two decades, but Darcian approximations are still common (and I used them frequently in my own work in the 1980s and 1990s).[4]

Clearly, many hydrological systems, above and belowground, evolve toward S&P configurations consisting of slow-flow, high-resistance components capable of storing or delaying flow and rapid-flow, low-resistance components capable of transmitting high water inputs and draining excess water. This occurs across the entire range of climates and geological settings on the planet. With respect to why, existing theories indicate that S&P configurations are maximum-efficiency patterns, giving a good reason for *why* they develop and persist, due to efficiency selection. This pointed to the main open research questions:[5]

- How do S&P configurations develop? Not the process mechanics of, for example, conduit enlargement, channel incision, or macropore formation, but rather, what system-level processes or mechanisms lead to S&P patterns?
- What (if any) underlying principles link the formation of S&P systems in surface watersheds, soils, groundwater, and channel-wetland complexes?

WHY STORE AND POUR?

S&P patterns and structures reflect the architecture and morphology of channel networks and drainage basins, of PFPs and matrix in soil, and of dual porosity in groundwater. These tend to emerge and persist because they *work* and because in many cases they are subject to growth and reinforcement by positive feedback. Arguments and theories based on optimality and efficiency, least action, maximum entropy production, thermodynamics, constructal theory, and percolation theory all point in this direction.[6]

To this we can add principles based on ecohydrology, the reciprocal interactions between hydrological processes and biota. Particularly in soils, S&P is highly beneficial to vegetation, allowing for H_2O storage between precipitation events and rapid export when excess water is received. Therefore, the dual-conductivity patterns that appear in a wide range of natural systems are likely linked to ecological processes. Plants and other biota have a direct role in creating macroporosity via root growth, burrowing, and tunneling and indirect roles owing to formation of soil aggregates and organic layers, pockets, or biomats. Biota are also crucial in invading and widening rock joints, fractures, and so on in chemical weathering, and vegetation has a well-known role in stabilizing channels.[7,8]

How Does Store and Pour Develop?

S&P configurations are favored by selection because they are generally efficient and offer ecological advantages. In a process mechanical sense, it can usually be determined how the preferential flow paths are formed, at least for specific landscapes. But we again return to the question of how Earth surface systems (ESS), without goal functions and lacking intentionality, achieve these configurations. I proposed a five-step emergent explanation, as shown in Figure 7.3. The steps are listed in order of logical necessity but do not always represent a temporal sequence. Each step reflects established, generally accepted principles.

1. Flows converge.
2. Convergent flows and preferential flow paths reinforced, enhanced, & locked in by positive feedback.
3. Flow paths intersect, forming networks.
4. Thresholds limit growth and extension of flow paths & networks, thereby maintaining seperation of contributing/storage areas (store) and flow paths (pour).
5. Store & Pour configurations are dynamically stable (thus tend to persist) & may be reinforced by ecohydrological feedbacks.

Figure 7.3. The emergence of store-and-pour patterns.

To describe the steps in more detail, first, concentrated flow happens. Concentrated flow is favored by principles of gradient and resistance selection and is more efficient for moving water than dispersed or diffuse paths. Concentrated pathways often develop opportunistically as water exploits preexisting routes such as microtopographic grooves or depressions, rock joints and fractures, roots and root channels, soil macropores, and so forth. Second, positive feedback enhances convergent flow paths. As water flows through soil pipes, macropores, conduits, and channels, debris and obstructions may be removed, reducing frictional resistance. Erosion—physical/chemical—may enlarge the features, and the wetting itself facilitates movement. Driving gravity gradients may be steepened, and development of these PFPs can also create local gradients that enhance the ability of the flow paths to capture water—for instance, water table drawdown adjacent to fluvial channels, lateral inflow to soil pipes, or suction exerted by plant roots. There is a major role for biota here in creating, maintaining, and enhancing several types of PFPs, but such features can also form abiotically. At the local scale, at least, the beginnings of S&P may arise simply due to segregation into concentrated flow paths and "other."

When converging flow paths intersect, they invariably combine to form networks (step three) in the down-gradient direction. These sometimes evolve maximum efficiency forms, such as dendritic fluvial networks. But even where flow paths are dictated by geological structures, the PFPs allow for greater overall flow velocities than if the entire system functioned as a single-porosity medium.[9] Emergence of these networks—also preserved by positive feedbacks—leads to S&P configurations at the network (or watershed or aquifer) scale.

Networks and channels or other PFPs cannot grow indefinitely; they are limited by thresholds (step four). Individual PFPs can only become so large due to mechanical limits on their maximum size (for instance, how steep stream banks can be or how large soil pipes can become before they collapse). Their size is also limited by the frequency and duration of flows capable of maintaining them. Minimum sizes are also limited, which constrains the ability of macropores or channels to extend upgradient and fill space. They cannot become infinitely small, and minimum sizes exist for conduits, channels, and macropores (and plant roots). A minimum runoff or flow contributing area is also necessary to provide maintenance flows. These thresholds prevent flow networks from becoming indefinitely dense and preserve the separation of rapid-flow components (PFPs) and slow-flow or storage components.

Step five holds that S&P patterns are usually dynamically stable, leading to their preservation via resistance selection or survival of the most stable. This is explored in more detail below. Ecohydrological feedbacks may also contribute to maintenance of S&P.[10]

SURVIVAL OF FLOW NETWORKS

For a landscape flow system to persist, it must experience limited degradation (e.g. infilling, structural collapse, vegetation clogging, desiccation) during dry, low-flow periods. It must also be able to efficiently shed excess water during wet periods, with some mechanisms for handling excess flow. Otherwise, the system can be destroyed or completely modified by erosional stripping, sedimentary burial, or persistent waterlogging. I'll use the terms *normal conveyance capacity* and *normal storage*, but I do not want to give the impression that there exists a clear distinction between normal and abnormal in landscapes. Rather, this is a shorthand whereby normal conveyance capacity refers to flux capacities of all perennial channels and conduits and macropores and other flow paths that are connected in unsaturated as well as saturated conditions. Normal storage refers to capillary storage in the soil matrix, matrix storage of groundwater, and surface water storage during non-flood and unsaturated conditions.

To limit breakdown in dry conditions, at least some PFPs must persist when flows are low. Dry time survival also requires some retention of water for use by biota and to maintain any fluvial, palustrine, or wetland features that happen to exist. Subsurface moisture storage occurs within the aquifer or soil matrix and, if moisture drawdown is not advanced, in subsurface cavities, conduits, and macropores. Surface water storage occurs in wetlands and ponded features such as ponds, lakes, sloughs, and non-flowing or slowly flowing subchannels. Significant long-term storage may also occur within vegetation.

Several possibilities exist for dealing with water inputs greater than normal transport capacities via fill-and-spill dynamics. *Spillways* have the same function as a dam spillway, activating a flow path for excess water. The most obvious spillways in natural hydrological systems are flood or high-flow channels on floodplains. Normally dry or non-flowing, these function as auxiliary channels that convey water downstream during floods. Floodplain rivers may also have distributary channels that are activated as stages approach banktops, conveying water into floodplain lakes, depressions, subchannels, and flood channels. In some low-gradient streams, subchannels are found that are essentially ponded by backwaters, with little or no downstream flow during normal inputs but which convey excess flow downstream when needed.

Many karst systems have spillways in the form of intermittent surface channels or high-flow conduits that are only activated during wet periods when subsurface cavities and normal-flow conduits are filled (Figure 7.4). Associated with these in some cases are perennial springs and spillway springs, often termed underflow and overflow springs. In soil hydrology, surface runoff, saturated throughflow (downslope flow within the soil), and percolation to groundwater serve as the main spillway mechanisms.

Some wetlands, such as alluvial bottomland swamps, also convey flow during floods. However, this slow, delayed flow (due to low slopes, shallow flow depths, and high roughness) is better viewed as a storage element in S&P systems.

Spilling may be two-stage process. Stage one is when surface depression storage (e.g., puddles) overflows, wet zones within soil enlarge and become connected, or when subsurface bedrock depressions fill up and overflow. This happens at or just beyond the upper end of normal flow or input conditions. Full spillway

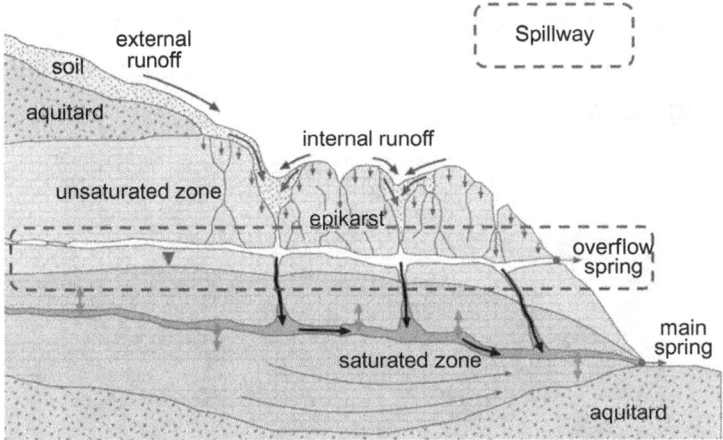

Figure 7.4. Karst spillway. During wet periods when the lower conduit is at capacity, the upper conduit and overflow spring are activated. Adapted with permission from A. N. Hartmann et al., "Karst Water Resources in a Changing World: Review of Hydrological Modeling Approaches, Review of Geophysics 52 (2014): 218–242, https:doi.org/10.1002/2013RG000443.

activation, stage two, occurs at above-normal flow conditions when additional pathways or processes of flow conveyance are activated.

When normal water storage and conveyance capacities are exceeded, S&P structures allow the water to be handled without any long-lasting or permanent transition. They also allow some water retention during dry, low-input periods.

In a high-input scenario, if the excess water is stored or delayed without triggering a large transformation, the system is sustainable and stable. Otherwise, transformative changes such as erosional stripping, sedimentary burial, or waterlogging occur, and the original system state is not maintained. Adjustments or adaptations of hydrologic, geomorphic, or ecological factors can in some cases move the system toward stability by increasing excess flow accommodation. For example, consider Figure 7.5 below. In this valley in the Ouachita Mountains of western Arkansas, water from an extreme precipitation event could not be removed rapidly enough, triggering a debris flow. This filled the valley, transforming it from a typical V-shaped mountain stream valley to the convex form you see, with the stream forced to the valley edge. So far, the new transformed system has been able to accommodate excess flows by storage in and subsurface flow through the deposits occupying the valley.

Figure 7.5. Valley in western Arkansas illustrating non-survival (transformation) of a system that was unable to handle very high inputs. See text for explanation.

In dry scenarios, hydrological systems with inadequate storage experience plant moisture stress (and plant mortality if the stress is prolonged or frequent). Unless some moisture storage capability is formed, this unstable non-sustainable situation leads to desiccation. If storage is adequate, vegetation can persist even with limited inputs (Figure 7.6). However, a system with high moisture storage capacity but limited ability to export flow ("all store, no pour") is likely to be, or become, a wetland environment with hydrophytic vegetation (except in arid climates) (Figure 7.7).

These scenarios illustrate how both store-and-pour capabilities are required for system survival. The box below gives some details of how the stability of hydrological systems can be determined. If you are not concerned with those details, here's the bottom line: the table below shows stability analyses of various scenarios (from Phillips, 2022, endnote 5). Under dry scenarios, stability depends on moisture storage, as limited storage capacity can lead to vegetation stress and mortality and desiccation. With water inputs greater than normal conveyance and storage capacities, dynamical instability occurs if high flows enlarge channels or other PFPs unless those expanding flow paths capture enough stored water.

Figure 7.6. This subtropical forest valley in southeast Queensland, Australia, can store sufficient moisture to survive dry periods and droughts.

Figure 7.7. Example of "all store, no pour": isolated bog wetland in Latvia. Photo by Martins Krastinsch via Pexels.com.

Inadequate spillway or secondary storage capacity also creates instability, while sufficient spillway conveyance for excess flows or adequate secondary storage capacity creates dynamical stability. The key lesson: survival of the flow system depends on dynamical stability, which in turn depends on moisture storage during dry spells and spillways for wet spells. Some illustrative examples follow.

Box 7.1

DYNAMICAL STABILITY OF FLOW SYSTEMS

Selection favors more stable configurations. But how can the dynamical stability of S&P patterns be assessed in a general sense, as opposed to examination of individual hydrological systems?

Here we take that on using a simple qualitative model of the interactions between storage of non-flowing water; slow, delayed flow; and flow and rapid movement of moisture through and out of the system (pour). In various situations, pour-and-store components may have negative or positive or negative effects on each other.

- If there is a fixed amount of excess water to be split among storage and pouring, the two are essentially competing for moisture and have negative effects on each other. Stored water or delayed flow reduces the amount in preferential flow paths and vice versa.

- In wet conditions, flow and storage can be mutually reinforcing (positive effects on each other). For instance, large storage quantities in the soil matrix or in aquifers may enhance preferential flows via return flow, saturated throughflow, exfiltration and saturation-excess runoff, while large channel or conduit flows may feed belowground storage. In some rivers, floodplain storage and spillway flow develop positive mutual effects during floods.

- Scenarios also occur where pouring negatively affects storage, which has positive effects on flow. This happens during drawdown of storage, as drainage of storage components feeds flow.

- Yet another scenario is when pour processes promote storage, which in turn limits flow. One example is when overbank flooding pours water into storage, which in turn limits channel flow. Another is where preferential flow paths are important for recharging aquifers and raising water tables, which in turn limits rapid flows.

- Pour/storage components may have self-limiting links (negative self-effects) where factors such as channel or conduit conveyance capacity or soil moisture storage capacity is a limiting factor.

- Pour/storage components may have self-reinforcing links (positive self-effects), as, for instance, when soil wetting increases hydraulic conductivity, facilitating faster recharge. Another example is when deeper flow depths inundate roughness elements, reducing resistance and speeding flow.

The figure below represents the relationships. The arrows could have various combinations of positive and negative signs under different situations as just described.

Mathematically, the stability of the system can be determined

$$a_{ps}a_{sp} - a_{pp}a_{ss} < 0$$

which is necessary and sufficient for a two-component system, The terms a_{ps}, a_{sp}, a_{pp}, a_{ss} signify feedback links from pour to store, from store to pour, self-effects on pour/flow components, and storage self-effects, respectively.

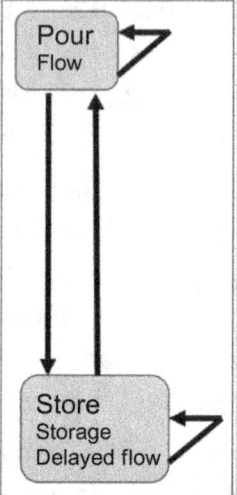

Box Figure 7.1. Store-and-Pour System Model

Where store-and-pour elements are mutually limiting, or mutually reinforcing, the system is dynamically *unstable* unless negative self-effects are dominant. If the signs of the interactions are opposite ($a_{ps}>0$ and $a_{sp}<0$, or $a_{ps}<0$ and $a_{sp}>0$), the system is dynamically *stable* unless there are positive self-effects that are both stronger than any negative self-effects and stronger than the pour-store interactions.

The table below shows stability analyses various scenarios (from Phillips, 2022; endnote 5). Under dry scenarios (1, 2), stability depends on moisture storage, consistent with the principle that limited storage capacity can lead to vegetation stress and mortality and desiccation. Under excessive H_2O inputs (greater than normal PFP conveyance and storage capacities as defined earlier) dynamical instability occurs when high flows enlarge channels or other PFPs, unless those expanding flow paths capture enough stored water (scenarios 5A, 5B). In scenario 6, insufficient spillway or secondary storage capacity also creates instability, while sufficient spillway conveyance for excess flows or adequate secondary storage capacity creates dynamical stability (scenarios 3, 4).

Scenario	Pour → Store (aps)	Store → Pour (asp)	Pour Self-Effects (app)	Store Self-Effects (ass)	Stability
1. Dry, with moisture storage	Flow enhances or has no effect on storage.	Retention of stored water reduces flow.	Negligible or negative	Negative due to increasing tension	S
2. Dry with limited moisture storage	Flow enhances or has no effect on storage.	Storage enhances flow due to limited retention, or no effect.	May be positive, negative, or negligible	Negligible, or negative due to increasing tension	U
3. Excessive moisture (<storage and conveyance capacities) with spillway overflow	Flow reduces storage.	Storage enhances flow by feeding spillways.	Negative due to finite conveyance capacity	Negative due to finite storage capacity	S
4. Excessive moisture (>storage and conveyance capacities) with secondary storage	Flow enhances storage by filling depressions, cavities, and gravity water.	Storage limits flow by activation of secondary storage.	Negative due to finite conveyance capacity	Positive, as saturation or filling activates secondary storage	CS[1]

Continued

Box Table 7.1. Continued

Scenario	Pour → Store (aps)	Store → Pour (asp)	Pour Self-Effects (app)	Store Self-Effects (ass)	Stability
5A. Excessive moisture with erosional enlargement of flow pathways	Flow reduces storage.	Storage reduces flow.	Positive due to increasing conveyance capacity	Negative due to finite storage capacity	U
5B. Excessive moisture with erosional enlargement of flow pathways	Flow reduces storage.	Storage enhances flow due to moisture capture by expanding channels, macropores, or conduits.	Positive due to increasing conveyance capacity	Negative due to finite storage capacity	CS[2]
6. Excessive moisture with limited spillway or secondary storage capacity	Positive or negligible.	May be positive, negative, or negligible.	Negative due top finite conveyance capacity	Positive	U

[1] Stable if $a_{pp} > a_{ss}$; unstable otherwise.
[2] Stable if a_{ps}, a_{sp} greater than self-effects, or if $a_{ss} > a_{pp}$; unstable otherwise.

EXAMPLES

The S&P conceptual model is designed to explain phenomena that are ubiquitous in nature, so examples are easy to find. The examples here are based on landscapes or systems I am familiar with and are intended to illustrate application of S&P reasoning. They are by no means a geographically balanced or comprehensive sampling.[11]

As a general example, consider forest soils and regoliths, which generally have PFPs operating in multiple directions through the subsurface. Event-scale moisture flux becomes imprinted in forest soils via self-organization of preferential flow systems (see the work of Roy Sidle, most recently at the University of Central Asia, and colleagues). In 2001, Sidle and coworkers found that the "backbone" for lateral flow paths is macropores formed by live and decayed roots, faunal burrows, piping erosion, and rock joints and fractures. Though individual macro pore segments are usually less than half a meter long, as sites get wetter they coalesce into larger preferential flow networks (Figure 7.8). Coalescence processes include flow through various types of macropores, fill-and-spill of small depressions in the underlying bedrock, flow in partings of weathered bedrock, exchange between macropores and smaller pores, and flow at the organic horizon-mineral soil interface (O-A horizon boundary) and in subsurface pockets of organic material and loose soil. Sidle et al.'s conceptual model of self-organization is based on activation of connectivity between macropore segments as soil moisture storage fills. Thus, normal conveyance capacity becomes better connected to provide spillway functions. Subsequent work confirmed the general applicability of the conceptual model in a variety of soils. Later work also showed how macropore connectivity during wet periods stimulates surface runoff—another spillway effect. As some of the other examples illustrate, this shows how storage features (in this case, the wettable nodes) begin storing some of the excess moisture inputs, eventually triggering or augmenting spillway processes.[12]

Figure 7.8. Soil in eastern North Carolina formed under subtropical forest. Segregated dark areas and visible partings show macropores, looser subsurface material, and buried organic matter pockets that become connected during wet periods.

Another example involves fluviokarst landscapes such as the Kentucky Inner Bluegrass example in Chapter 6. Fluviokarst landscapes are characterized by complex, interconnected combinations of ground and surface water flow, underground conduits and cavities, and surface channels.

The general nature of many fluviokarst flow systems has remained intact despite climate and other environmental changes, human impacts, and transitions within the landscapes among karst and fluvial flow regimes and landforms. This implies general, broad-scale stability at the regional scale and (according to the conceptual model above) that the necessary conveyance and storage capacities and mechanisms are present.[13]

Perennial (surface) stream channels, groundwater conduits, and surface-ground water interconnections (i.e., sinks or swallets and perennial underflow springs) provide normal flow conveyance. Subsurface karst cavities, vertical or near vertical or unconnected rock joints, epikarst (weathered upper limestone layer), and soil provide normal moisture storage, which is sufficient to maintain the system through dry periods. When moisture inputs are excessive, secondary storage is achieved by filling larger subsurface cavities, depressions in the upper epikarst (often bowl or saucer-like weathering cavities on surface rock outcrops or beneath a thin soil cover), and karst surface depressions such as dolines. Spillways are activated as filled karst conduits or cavities spill into overlying, normally dry stream channels and overflow springs begin flowing.

For a third example of why and how S&P systems work, consider woody vegetation invasions of drylands, which provide opportunities to observe how vegetation change re-engineers hydrological systems. When shrubs invade semi-arid environments, the result is often landscape divergence: vegetated patches or thickets with thicker soil and more soil moisture interspersed with bare or sparsely vegetated areas. The former become islands of soil fertility, moisture, organic matter, and vegetation in a sea of unvegetated, low-fertility, dryer areas (Figure 7.9).[14]

Figure 7.9. Matrix of vegetated shrub patches and bare ground, Tikaboo Valley, Nevada. From a photo by Chris Smith via Flickr.com under the CC-BY 2.0 license.

PFPs are created or enhanced as woody plants become established, funneling water into the soil. In the vegetated patches, normal conveyance feeds storage, which in turn supports plant survival through the inevitable dry periods. When inputs are excessive (for example, during convective thunderstorms), additional storage is limited. This is compensated by spillway mechanisms (runoff) exporting water to dry patches. The latter provide landscape-scale spillway capacity, but their lack of moisture storage capability inhibits vegetation establishment and survival.[15]

Tidal marshes are a much different type of example—always wet to varying extents, with bidirectional fluxes, and regular, frequent ebb and flood tide pulses and reversals. The main water inputs and outputs are due to tidal and storm flooding. The *tidal prism* is the volume of water associated with the difference between low and high tides. As sea level rises (as it has been doing in most areas throughout the Holocene; accelerating in many at present), tidal prisms increase (unless marsh surface accretion keeps pace with relative sea level rise and the marshes maintain their surface area; on average globally, this is not the case).

According to the S&P concept, marsh survival hinges on auxiliary storage and/or spillway capacity (Figure 7.10). Accretion may increase matrix moisture storage capacity, and secondary storage is linked to formation of depressional storage such as marsh ponds and salt pans. However, these features may also accelerate marsh fragmentation. Conveyance capacity may be increased by expansion of tidal networks as tidal prisms increase. Additional spillway capacity could be increased by high-flow channels not inundated during normal tides but activated during higher tides or storm inundation.[16]

Net marsh loss that often occurs in response to coastal submergence can be seen through the S&P lens as evidence of failure to build secondary storage and spillway capacity rapidly enough. The geomorphic transformation of some types of brackish marshes and of freshwater swamps adjacent to some estuaries in North Carolina during an extreme storm surge event can similarly be attributed to the storage and spillway capacities of the wetlands being overwhelmed.[17]

In the final example, we address human modifications of hydrology: artificially drained farmland in eastern North Carolina. The outer coastal plain of North Carolina is low (<6 m above sea level) and flat, with high water tables and frequently saturated soils. Corporate-scale agriculture and intensive silviculture require that water tables be lowered by artificial drainage, typically

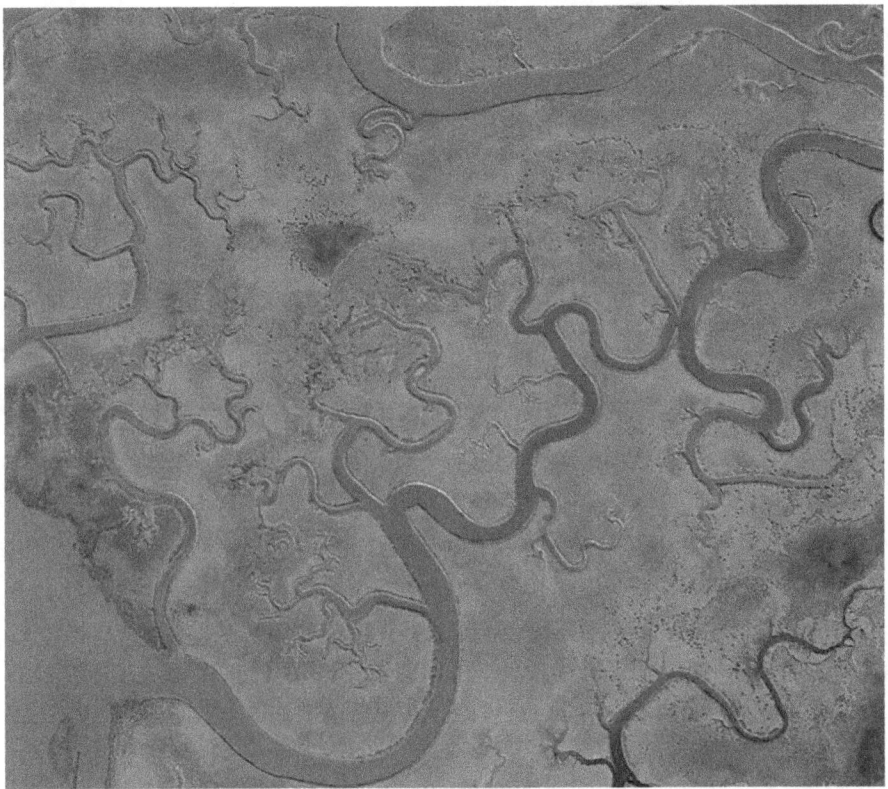

Figure 7.10. Tidal marsh at Virginia Coast Reserve showing tidal channel networks, marsh depressions, and high flow channels. Area shown is about 0.62 × 0.65 km. Google Earth™ image.

in the form of rectangular networks of drainage ditches and canals. Before-and-after comparisons between undrained, drained, and post-drained conditions can be made based on studies of the hydrological impacts of ditch and canal systems and of the hydrology and geomorphology of the artificial channels themselves.[18]

Normal storage in the undrained condition is entirely within the soil and underlying unconsolidated sediments, with water tables regularly at or within 0.5 m of the surface. Shallow surface depressions provide secondary storage conditions, along with local ponding and the rise of water tables above the ground surface. Flow conveyance capacities are low due to minimal topographic gradients and are limited to low-gradient streams that are subject to backwater effects. The unmodified landscape is a mosaic of wetlands and other poorly drained lowland environments supporting vegetation tolerant of soil saturation.

Ditches and canals increase conveyance capacity and shift the role of water table rise to or near the surface from a normal storage mechanism to a means of temporary secondary storage. Spillway mechanisms are still limited by low gradients but sometimes augmented with pumping systems. Water levels are also often mechanically manipulated via water control structures, creating an engineered S&P system.

Canal and ditch flows are not adequate to maintain the channels, however, as drainage density has been increased so that the contributing area is not sufficient, given low velocities due to very low slopes. The artificial channels therefore rapidly lose conveyance capacity without maintenance, such as vegetation and plant debris removal, and re-excavation (maintenance is generally needed every two to five years).

Post-drainage, after maintenance is discontinued and/or water control structures such as flashboard risers are permanently left in place to dam canals, the channels are essentially converted to linear ponds. This is like the pre-drainage state, with the ditches and canals augmenting storage. Passive wetland restoration by non-maintenance or simple plugging of canal outlets is often viable. The main difference between pre- and post-drainage is that the channels may fill and spill during exceptionally wet periods to provide some limited spillway capacity.

STORE AND POUR AS EFFICIENCY SELECTION REINVENTED

Hydrologic flow patterns in landscapes often take forms that approach or achieve optimality with respect to flux efficiency, thermodynamics, or biological needs. But like other ESS, they do this in the absence of any intentionality or goals. Additionally, notwithstanding some inheritance in the form of historical contingency and relic features, those optimal configurations are not heritable or transferrable. Subsequent flow systems on the mutable landscape must develop analogous structures independently. DNA allows organisms to inherit and "remember" environmental adaptations, but S&P and other abiotic adaptations require perpetual reinvention.

New surfaces and landscape substrates are constantly being created by sediment deposition, sea and lake level changes, glacial retreats and advances, impact events, uplift, and volcanic eruptions. Existing ones are constantly being destroyed or renewed by erosional stripping, earthquakes, mining, construction, warfare, fire, and cataclysmic storms. The flow systems that develop on these new surfaces or after clock-resetting events have no programming or genetic memory to evolve S&P structures, yet more often than not (remember, selection is probabilistic, not deterministic) S&P re-emerges—or, to put it anthropomorphically, is reinvented.

ADAPTATION AND CONTINGENT PARTITIONING

Does the development of S&P structures and the associated partitioning of water among normal flow and storage, spillway flow, and secondary storage constitute evolutionary adaptation? To address this, we'll expand a bit from considering store vs. pour within the system to add consideration of water inputs to and output from a hydrological system. We'll also shift our focus from the factors that promote survivability of a hydrological flow system over a range of high- and low-input conditions, as examined above, to how systems adapt to changing wet and dry states from one event or episode to the next.[19]

Adaptation

Landscapes certainly *respond* to changes in boundary conditions, energy and matter inputs, and to internal processes and structures. But response and adaptation are not the same. You will *respond* to getting hit in the head with a hammer but not *adapt* to it.

Landscape responses are always finite and often decelerate—i.e., quick response immediately after a change or disturbance, with the response decreasing to negligible rates, often relatively rapidly. In this sense, they can be said to have *adjusted*. Adaptation in biology is often defined as the adaptation of living things to environmental factors for the ultimate purpose of survival, reproduction, and optimal functioning. This suffers from the disadvantage of defining something as itself (biological adaptation is the adaptation of living things. . . .). What if you substitute environmental systems or landscapes for living things and adjustment for adaptation? Then *adaptation is adjustment of environmental systems to enable survival and efficient functioning*. If we accept this definition, we can legitimately consider landscape adaptation. If hydrological (or other Earth surface) systems adjust to enhance their persistence or functioning, they can be said to adapt.

A search of any scholarly database will turn up many titles of the form "_____ as a Complex Adaptive System." Environmental (including hydrological) systems, along with social, political, cultural, biomedical, and computing systems have been labelled *complex adaptive systems* (CAS). The CAS terminology and framework as applied to hydrology, water resources, and environmental sciences has focused on management or design such as stormwater control or natural resource management systems, on global-scale phenomena such as the biosphere or the entire planet as a CAS, or on CAS as a metaphorical or descriptive tool. In contrast to the CAS literature, we focus on whether and how hydrological systems (as a representation of landscapes) adapt as defined above, which is at least broadly analogous to biological adaptation.

Partitioning

Budgeting or partitioning of water is fundamental to hydrology. The hydroclimatic water balance, for instance, is typically one of the first topics encountered in hydrology texts or in the water chapters of ecology, meteorology, or physical geography texts and is fundamental to applied hydrology and water resource management. The water balance or budget analyzes the partitioning of precipitation inputs to evapotranspiration, soil moisture storage, and runoff (including percolation to groundwater). Rainfall-runoff analysis is another example, where effective precipitation (precipitation minus evapotranspiration) is allocated to surface runoff, infiltration, soil moisture storage, saturated throughflow, and groundwater.

These and similar approaches are applied to dynamic situations at event to seasonal scales and recognize that partitioning varies between episodes or events. Moreover, the variable partitioning applies not only quantitatively but also qualitatively. The latter reflects variations in the operation—or not—of pathways or partitions. Soil moisture recharge shuts down when soil is saturated, for example, and surface runoff does not occur if soil moisture is not fully recharged or if precipitation intensity is less than infiltration rates. Moreover, some flow pathways and connections can be reversed. Tidally influenced systems are an example where reversal is frequent and normal, but it also happens in other hydrological situations, such as infiltration of water into vs. exfiltration out of soil.

Partitioning of water is therefore *contingent* on synoptic situations. Does dynamic, contingent partitioning of water constitute adaptation? A hydrological system is "successful" if it survives (and maybe even grows) through (sub-catastrophic) environmental changes; efficiently processes water via storage, utilization, and transport; and supports a biological community. So, if the system adjusts in such a way as to become or remain successful, it is adaptive.

Geophysical and Ecological Jobs

What is hydrological system's "job"? If more precipitation falls than
the ground can hold or organisms can use, it must go somewhere,
driven by gravity. Associated processes such as erosion, sediment
and solute transport, and biogeochemical fluxes are byproducts of
the fundamental hydrological job of accommodating excess water.
From a geophysical perspective, the job of a hydrological system
is simply to move excess water.

However, hydrological systems also have an ecological job: to
support biota, particularly plants, by supplying water, removing
or transferring byproducts, and enabling biogeochemical cycling.
Organisms are obliged to adapt to hydrological conditions, particularly those dictated by climate. But through their own impacts
on the hydrological cycle and various ecosystem engineering processes, plants and other organisms affect, as well as are affected
by, hydrology. To be successful, hydrological systems must do both
the geophysical and ecological jobs. In many cases, the ecological
functions are stressed when inputs are low and the geophysical
functions when inputs are high. With low inputs, ecological functions may not be fully or optimally supported, and the geophysical
job is irrelevant; there is no excess water to move. If the system
can respond quickly to increased inputs, that is advantageous to
the ecological job (Figure 7.11).

Water supply for ecological functions is not limited during excess inputs, and the geophysical job becomes primary (anaerobic
conditions and other impacts of excess moisture can cause ecological stress, but these will be minimized if the geophysical role
is performed). To survive, mechanisms to handle excess water
from floods, downpours, and so on must exist. This happens via
S&P with augmented storage and, especially, high-input-activated
spillways.

This takes us back to the question of dynamical stability. To
perform the geophysical job during wet periods, a hydrological

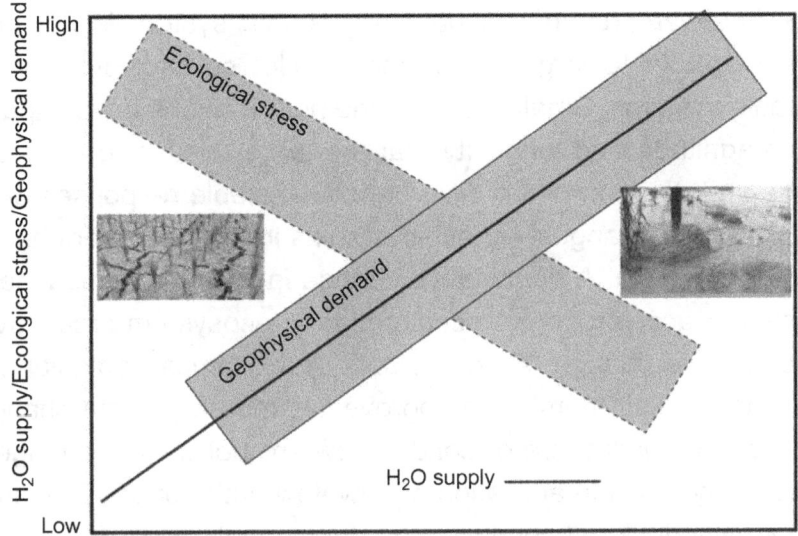

Figure 7.11. Highly simplified conceptual model of the relationship between water inputs to a flow system, stress on the ecological functions, and the need to move excess water (geophysical demand). Ecological stress and geophysical demand are shown as bands to represent the fact that they will vary according to factors other than moisture supply. All are shown as straight lines or bands for simplicity; actual functions are more complex (for instance, in some systems ecological functions may be limited by excess moisture).

system should be able to maintain itself in response to changes and disturbances,[20] which means it should be dynamically stable. However, to perform its ecological job, the system should *not* maintain or restore the state it is in during dry conditions—rather, it should be able to quickly respond to new inputs and move to a new state. That is, it should be dynamically unstable.

Let's take a closer look at dynamical (in)stability and its relationship to the geophysical and ecological jobs and to adaptation.

ADAPTABILITY AND DYNAMICAL STABILITY

A dynamically stable system is insensitive to small perturbations or changes and can maintain itself or return to its pre-disturbance state afterward (even a dynamically stable system is not necessarily stable to large disturbances). Not so with dynamically unstable systems. Small changes and perturbations have impacts with magnitude and longevity that are large relative to the magnitude and duration of the disturbance. Unstable responses often result in state changes—qualitative shifts in the nature or character of the system. A quantitative change in relative abundance of species or acceleration or deceleration of ecosystem productivity is not a state change. A regime shift—for example, grassland to shrubland or salt marsh to mangrove swamp—is a state change. Increases in macropore or conduit flow are not a state change—unless they result in activation of new flow paths or storage, such as normally dry karst conduits or surface ponding.

Dynamical stability is consistent with some concepts and definitions of resilience and has sometimes been conflated with resilience.[21] But system state transitions triggered by dynamical instability are sometimes necessary to adapt to changes.

Simple examples are when a stream shifts from single-channel to multichannel flow in response to floods, normally dry conduits in karst are activated during wet episodes, or plant stomatal resistance changes during dry periods. Adaptation can therefore be associated with either dynamical stability or instability.

We tend to think of stability as either/or, though it is hardly constant in space, time, or at different scales. That is, we often see a system as either stable, unstable, or on a cusp between the two. This is often valid enough, particularly on longer timescales or with respect to ESS survival and persistence across a range of inputs and external environmental conditions. In ESS, however—characterized here by hydrologic systems—dynamical stability is situational. At more detailed temporal or spatial scales, systems that are stable and persistent at a broader scale (e.g., S&P systems with adequate storage and spillways) may be stable or unstable. To use a human analogy, short-term social or political instability that results in beneficial changes may contribute (or even be necessary) to the long-term stability of a society.

Mathematically, dynamical stability is a function of the negative and positive relationships among system components. But in hydrology and landscapes, those relationships may vary between positive, negative, and zero (negligible effects) during wetter or drier episodes, thus potentially shifting stability status. This has some parallels in other ESS. Over landscape evolution timescales, geomorphological systems may undergo shifts between dynamically stable and unstable modes as interrelationships increase or decrease in relative strength or intensity.[22] Ecological systems may shift between positive and negative self-effects of density (e.g., seed source effects, facilitation vs. competition, resource limitation) during succession or following disturbances.

First, we'll examine the dynamical stability properties of a generalized hydrological system and the ability of the system to adapt to changes in its external environment and flow and storage dynamics within the system. Then, as usual, we'll look at some specific examples.

Generalized Flow System

The generalized hydrologic system in Figure 7.12 is applicable in a broad, first-order sense to most flow systems. It has just four components: inputs of water, storage within and flow through the system, and outputs. These components can be expanded for more specific cases. A hillslope hydrological system, for instance, has precipitation and run-on inputs; storage in the regolith matrix and macropores (and possibly in groundwater cavities or aquifers and surface depressions); flow through it, including overland runoff, percolation, and lateral subsurface flows; and outputs of evapotranspiration and various types of outflow.

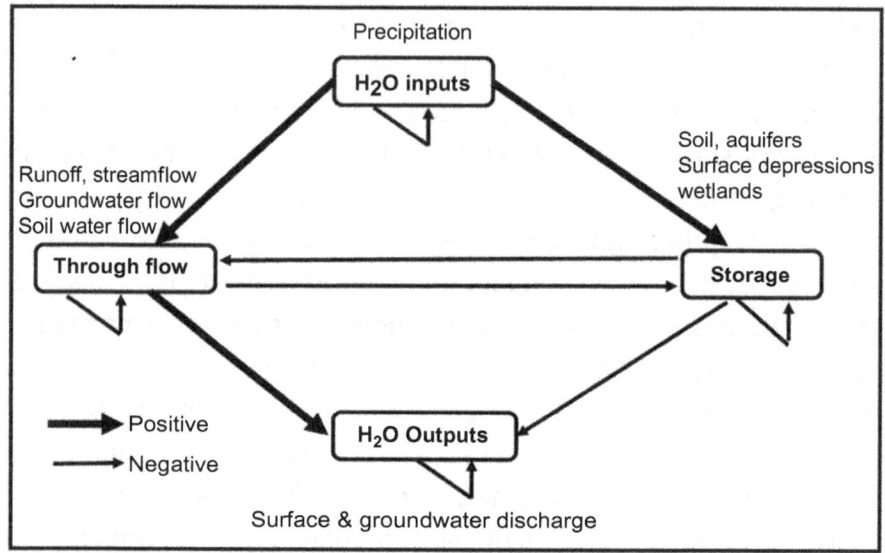

Figure 7.12. Generalized hydrological system. Throughflow refers to all water flux through the system, as opposed to throughflow within soil.

The thicker positive arrows show flow from one component to another; the thinner negative ones indicate that one component is "competing" for water with the other in a zero-sum manner (that is, a partitioning). Figure 7.12 suggests, for instance, that the more water that goes into storage, the less there is allocated to flow through and out of the system. The figure shows all self-effects as negative (self-limiting), but they may sometimes be positive, as when wetting, or drowning of roughness elements, enhances conveyance capacity. Table 7.1 shows a generalized interaction matrix for the system. By analyzing this under different scenarios of positive, negative, or zero values, stability in different situations can be determined. The interaction matrix shows some entries for links not shown in Figure 7.12—these are relationships that are not necessarily typically present but do occur in some scenarios. Table 7.2 explains the different signs the arrows in Figure 7.12 or the a_{ij} in Table 7.1 might take in different systems or scenarios. As a compromise between showing my work and losing readers by burdening the narrative

Table 7.1. INTERACTION MATRIX FOR GENERALIZED HYDROLOGICAL SYSTEM

	Inputs	Throughflow	Storage	Outputs
Inputs	a_{11}	a_{12}	a_{13}	0
Throughflow	0	a_{22}	a_{23}	a_{24}
Storage	0	a_{32}	a_{33}	a_{34}
Outputs	0	a_{42}	0	a_{44}

Nonzero entries may be positive, negative, or negligible (≈ 0) under various circumstances.

with mathematics and other technicalities, I have banished the stability analysis methods to endnotes.[23]

Scenarios

In addition to the base case shown in Figure 7.12, I examined the stability of three different types of hydrological systems under different circumstances. This involves choosing the appropriate signs for each link or interaction matrix entry and then doing the math to determine whether, or under what conditions, the mathematical criteria for dynamical stability are met. For each scenario, Table 7.3 shows one of five outcomes. Stable or unstable means that system configuration under that scenario is either unequivocally (un)stable no matter what or that (in)stability would require relative values of interactions that are hydrologically unrealistic for the case in question. For example, under the dry soil scenario, it is mathematically possible for the stability criteria to be met, but it would require self-limitations on soil moisture flow and storage to dominate the system, which ain't happening in the dry scenario. Generally (un)stable means that is mathematically and hydrologically possible for the other outcome to occur in real systems, but rare and unlikely. Conditionally stable indicates that dynamical stability or instability could plausibly occur, depending on circumstances.

Table 7.2. DESCRIPTION OF HYDROLOGICAL FLOW SYSTEM MODEL LINKS

Link	Description	Sign(s)
a_{11}	Water input self-effects	Negative in most cases as H_2O supply depleted. Zero for small events or during early stages of inputs.
a_{22}	Throughflow self-effects	Negative at high flow, limited by conveyance capacity. Potentially positive when wetting, flushing, inundation. Decrease resistance. Negligible or zero at moderate flows.
a_{33}	Storage self-limitations	Negative as storage capacities approached or exceeded. Potentially positive during wetting front propagation or as storage elements become connected. Negligible or zero at moderate storage.
a_{44}	Outflow self-effects	Negative as flux capacities approached or exceeded. Zero otherwise.
a_{12}	Inflow effects on throughflow	Usually positive, except. . . . Zero during soil or groundwater wetting.
a_{13}	Inflow effects on storage	Usually positive, except. . . . Zero at saturation.
a_{21}	Throughflow effects on input	Negligible or zero in most cases[1]

Link	Description	Sign(s)
a_{23}	Throughflow effects on storage	Negative in many cases due to competitive partitioning of inputs. Potentially positive at high flows due to flow diversions to storage, activation of distributaries.
a_{24}	Throughflow effects on outflow	Positive.
a_{31}	Storage effects on input	Negligible or zero in most cases.[1]
a_{32}	Storage effects on throughflow	Negative when zero-sum partitioning of inputs occurs. Positive if storage is draining via throughflow.
a_{34}	Storage effects on outflow	Negative.
a_{42}	Outflow effects on throughflow	Negative if slow outflow reduces gradients or blocks throughflow. Positive if accelerated outflow increases gradients. Zero in most cases.

[1] Here I assume inputs are externally controlled. In some cases, at broad scales where local moisture recycling is important, these feedbacks could be nonzero.

It is indeed important to balance the mathematical and hydrological reasoning. As mathematicians Ian Stewart and Martin Golubitsky put it in their book *Fearful Symmetry*, "Mathematically, you can balance a chain end to end, provided each link sits precisely above the one below it. . . . This is impossible in the real world. So mathematical *existence* of solutions to equations isn't enough."

These results show that hydrological flow systems tend to be dynamically unstable during dry or recharge conditions, thereby facilitating state changes when new moisture inputs arrive. When

Table 7.3. STABILITY ANALYSIS RESULTS FOR HYDROLOGICAL SCENARIOS.

Scenario	Description	Stability
Base	Generalized system shown in Figure 7.12.	Conditionally stable
Soil	Applies to a soil pedon.	
Soil: dry	Dry conditions when moisture storage is being depleted.	Unstable
Soil recharge	New inputs to previously dry soil.	Generally unstable
Soil wetting	Soil approaching limits to storage and normal flow through.	Generally stable
Soil saturation	Normal storage, flow through and outflow limits reached.	Stable
Soil: infiltration excess	Infiltration-excess or Hortonian overland flow is a transient phenomenon that occurs when precip- itation inputs exceed infiltration capacity, producing surface runoff.	Conditionally stable

Scenario	Description	Stability
River channel	Applies to a reach of a perennial alluvial fluvial channel.	
River channel: flow connectivity	Minimum discharge necessary to maintain downstream flow.	Unstable
River channel: bed inundation	Threshold discharge to cover entire channel bed.	Unstable
River channel: in banks	River stages higher than bed inundation; lower than channel-floodplain connectivity.	Conditionally stable
River channel: channel floodplain connectivity	Flow between channel and floodplain activated via distributaries, crevasses, or overbank stage.	Stable
River channel: flood	Spillways and secondary storage activated.	Stable
Fluviokarst	Applies to karst systems with interactions between surface and subsurface flow; with karst and fluvial processes controlled by the same base level.	

Continued

Table 7.3. Continued

Scenario	Description	Stability
Fluviokarst: dry	Dry conditions: moisture storage depleted.	Unstable
Fluviokarst recharge	New inputs to pre-viously drying system.	Unstable
Fluviokarst: saturation	Soil and epikarst are saturated; under-ground cavities and conduits approaching capacity.	Conditionally stable
Fluviokarst: overflow	Filled conduits and cavities spill into nor-mally dry cavities and stream valleys; over-flow conduits and springs activated.	Stable
Fluviokarst: flood	Same as overflow except damming and backwater effects limit flow through.	Stable

there is excess moisture, they are generally dynamically stable, allowing the systems to be maintained. Adaptation is sometimes linked to dynamical instability when state changes are necessary or advantageous. Adaptation can also be associated with sta-bility, where resilience is necessary for persistence or survival. Adaptive systems may have both stable and unstable modes in various circumstances or situations.

Figure 7.13 is a simplified summary showing that demands on the geophysical job of the flow system (moving or storing excess water) are directly related to water inputs, while demands on the ecological job are inversely related. With low inputs, dynamical instability allows quick state changes in response to new inputs to support the ecological job. Stability during wet periods maintains the system so that it can do its geophysical job. Transitions between the ecological and geophysical regimes are connected to changes in the relative importance of flow and storage feedbacks and activation of flow paths and auxiliary storage (or deactivation in the case of declining inputs). Hydrological systems that exhibit these dynamics are indeed adaptive.

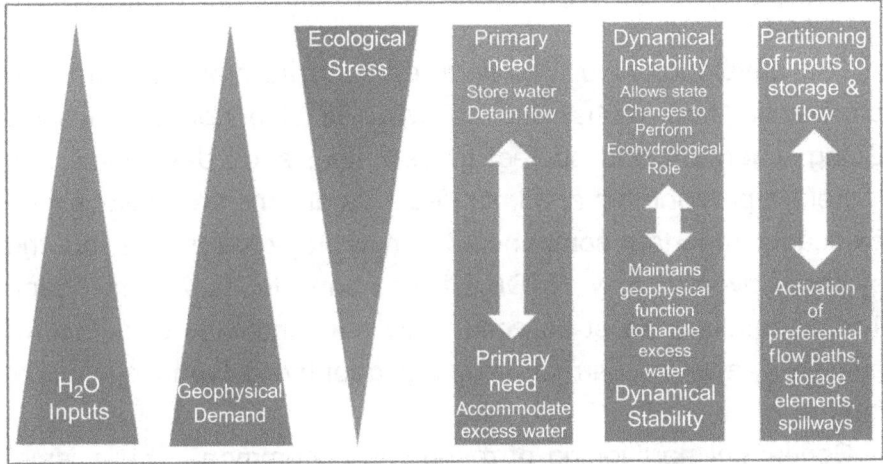

Figure 7.13. Simplified summary of relationships among water inputs, ecological stress, geophysical demand to move excess moisture, dynamical stability, and adaptations.

I am not aware of other studies explicitly linking hydrological responses to evolutionary adaptation, but certainly examples of contingent partitioning and switches in flux-storage relationships are not hard to find.[24]

We have not considered all plausible combinations of interactions and self-effects among inputs, throughflow, storage, and output. Yet, the scenarios are diverse enough and common enough to illustrate the fact that partitioning of water inputs within hydrologic systems is strongly conditional even in a qualitative sense, and contingent on synoptic situations. Of sixteen possible links among system components, only four are always zero. Of the twelve that are always or sometimes nonzero, nine may have at least two different signs (negative, zero, positive).

Environmental adaptation in this case is linked to the performance of the ecological and geophysical jobs. Non-adapted hydrological systems inevitably exist, as disturbances and changes may delay, disrupt, or even prevent evolutionary adaptation. Development of key mechanisms such as spillways, PFPs, and secondary storage is not preordained (no laws dictate it), and soil, rock, and water have no goals or desires. The adaptation, where and when it happens, is emergent and subject to selection.

The development of key features such as storage elements, preferential flow paths, and spillways is also nondeterministic. Being emergent and subject to selection, such development is therefore probabilistic and imperfect. Organisms are obviously inextricably important components of hydrological systems, but the systems overall carry no DNA instructions to help them adapt. The adaptations must be continually reinvented by selection for efficiency and for carrying out the geophysical and ecological jobs.

Contingent partitioning of resources is common in organisms, as they allocate water, carbohydrates, nutrients, and so forth to meet various needs and respond to different supply-demand situations. Recent research on communication and transfers among plants using their mycorrhizal networks suggests that such contingent partitioning can also occur at the population, community, and ecosystem levels.[25] And the hydrological example shows that it can occur in geophysical phenomena independently of biological controls. Reinventive adaptation is, perhaps, not so mysterious after all.

NOTES

1. See the preferential flow section of Chapter 6.
2. Some key references influential in my thinking: H. Lin, "Linking Principles of Soil Formation and Flow Regimes," Journal of Hydrology 393 (2010): 3–19; H. H. G. Savenije and M. Hrachowitz, "HESS Opinions: 'Catchments as Meta-Organisms'—A New Blueprint for Hydrological Modeling," *Hydrology & Earth System Sciences* 21 (2017): 1107–1116.
3. A. G. Hunt, "Spatio-Temporal Scaling of Vegetation Growth and Soil Formation from Percolation Theory," *Vadose Zone Journal* 15 (2016): https://doi.org/10.2136/vzj2015.01.0013; A. G. Hunt, "Use of Constructal Theory in Modeling in the Geosciences," in *Fractals. Concepts and Applications in Geosciences*, edited by B. Ghanbarian and A. G. Hunt (Boca Raton, FL: CRC Press, 2017), Ch. 12, p. 15; A. G. Hunt and B. Ghanbarian, "Percolation Theory for Solute Transport in Porous Media: Geochemistry, Geomorphology, and Carbon Cycling," *Water Resources Research* 52 (2016): 7444–7459, https://doi.org/10.1002/2016WR019289.
4. This type of flow involves a mass of water moving through a porous medium; water soaking into (or dripping out of) a sponge is an example. It gets its name from Darcy's law, an equation formulated by French engineer Henry Darcy in 1856, describing flow through porous media (such as an aquifer) as a function of the difference in hydraulic head and the hydraulic conductivity of the medium.
5. These ideas were published in J. D. Phillips, "Store and Pour: The Evolution of Flow Systems in Landscapes," *Catena* 216 (2022): 106,357.
6. These theories and explanations are mutually consistent, and there is considerable overlap.
7. Zehe et al. (2013, 4318) speculated that "co-evolution selects species whose ecological optimum (pattern and density) coincides with the thermodynamic optimum partitioning of rainfall water" into overland flow and infiltration in such a landscape. These optimum configurations may be most probable states in landscape evolution. E. Zehe et al., "A Thermodynamic Approach to Link Self-Organization, Preferential Flow and Rainfall-Runoff Behavior," *Hydrology and Earth System Sciences* 17 (2013): 4297–4322. See also Lin (2010) and Savenije and Hrachowitz (2017, endnote 2).
8. Recently several teams have presented evidence that "vegetation optimality" explains global trends in relationships between evapotranspiration and stream flow. The Vegetation Optimality Model (VOM) optimizes plant properties and behavior such as rooting depths, vegetation cover, and stomatal resistance to maximize the difference between total carbon taken up from the atmosphere and the C used for maintenance of plant tissues (the long-term net carbon profit). The VOM is proposed to explain the fact that many watersheds are close to the physical limit set by conservation of mass in the relationships among precipitation, evapotranspiration,

and net radiation (R. C. Nijzink and S. J. Schymanski, "Vegetation Optimality Explains the Convergence of Catchments on the Budyko Curve," *Hydrology and Earth System Sciences* 26 (2022): 6289–6309). Another group (A. G. Hunt, M. Sahimi, and B. Ghanbarian, "Predicting Streamflow Elasticity Based on Percolation Theory and Ecological Optimality," *AGU Advances* 4 (2023): https://doi.org/10.1029/2022AV000867) reached similar conclusions based on optimizing net primary productivity of vegetation. While the first study focused on carbon dynamics, the second emphasized the transition from water to energy limitations in evapotranspiration. However, another study showed that climate change and disturbances can decouple vegetation, evapotranspiration, and streamflow by altering the size, availability, and connectivity of source pools for ET and discharge (N. G. McDowell et al., "Ecohydrological Decoupling under Changing Disturbances and Climate," *One Earth* 6 (2023): 251–266). Indeed, the Nijzink and Schymanski and the Hunt et al. analyses assume that changes in storage are negligible. Even more provocative is a recent proposal that soils and their hydraulic properties are "overrated" in hydrology, because "the terrestrial ecosystem manipulates the soil to satisfy specific water management strategies. . . ." (H. Gao, F. Fenicia, and H. H. G. Savenjie, "HESS Opinions: Are Soils Overrated in Hydrology?" *Hydrology and Earth System Sciences* 27 (2023): 2607–2620).

9. Worthington provides some convincing evidence on this: S. R. H. Worthington, "How Preferential Flow Delivers Pre-Event Groundwater Rapid to Streams," *Hydrological Processes* 33 (2019): 2373–2380; S. R. H. Worthington, G. J. Davies, and E. C. Alexander, Jr., "Enhancement of Bedrock Permeability by Weathering," *Earth-Science Reviews* 160 (2016): 188–202; S. R. H. Worthington and D. C. Ford, "Self-Organized Permeability in Carbonate Aquifers," *Ground Water* 47 (2009): 326–336.

10. Some hydrologists emphasize these ecohydrological controls and feedbacks; see references in endnotes 7 and 8.

11. These are presented in more detail in Phillips, 2022 (endnote 4).

12. On multidirectional preferential flow paths, see J. D. Phillips, L. Pawlik, and P. Samonil, "Weathering Fronts," *Earth-Science Reviews* 198 (2019): 102295; P. Šamonil et al., "Soil, Regolith, and Weathered Rock: Theoretical Concepts and Evolution in Old-Growth Temperate Forests, Central Europe," *Geoderma* 368 (2020): 114261.

    The Sidle et al. conceptual model is described in R. C. Sidle, S. Noguchi, Y. Tsuboyama, and K. Laursen, "A Conceptual Model of Preferential Flow Systems in Forested Hillslopes: Evidence of Self-Organization," *Hydrological Processes* 15 (2001): 1675–1692.

    Some important examples of the subsequent work mentioned include H. Liu and H. Lin, "Frequency and Control of Subsurface Preferential Flow: From Pedon to Catchment Scales," *Soil Science Society of America Journal* 79 (2015): 362–377; J. L. Nieber, T. S. Steenhuis, T. Walter, and M.

Bakker, "Enhancement of Seepage and Lateral Preferential Flow by Bio-pores on Hillslopes," *Biologia* 61 (2006): 225–228; J. L. Nieber and R. C. Sidle, "How Do Disconnected Macropores in Sloping Soils Facilitate Prefer-ential Flow?" *Hydrological Processes* 24 (2010): 1582–1594; G. V. Wilson et al., "Hydrologic Connectivity and Threshold Behavior of Hillslopes with Fragipans and Soil Pipe Networks," *Hydrological Processes* 31 (2017): 2477–2496.

13. J. D. Phillips, "Landform Transitions in a Fluviokarst Landscape," *Zeitschrift für Geomorphologie* 61, (2017): 109–122; J. D. Phillips, "Historical Contin-gency in Fluviokarst Landscape Evolution," *Geomorphology* 303 (2018): 41–52.

14. I have done a bit of work on this phenomenon, but I rely mainly on ex-perimental studies from the Chihuahuan Desert of North America by Tony Parsons, Athol Abrahams, John Wainwright, and William Schlesinger: A.D. Abrahams, A. J. Parsons, and J. Wainwright, "Deposition of Rainwater under Creosotebush," *Hydrological Processes* 17 (2003): 2555–2566; M. Neave, A. D. Abrahams, "Vegetation Influences on Water Yields from Grassland and Shrubland Ecosystems in the Chihuahuan Desert," *Earth Surface Processes and Landforms* 27 (2022): 1011–1020. A. J. Parsons, J. Wainwright, W. H. Schlesinger, and A. D. Abrahams, "The Role of Overland Flow in Sediment and Nitrogen Budgets of Mesquite Dunefields, Southern New Mexico," *Journal of Arid Environments* 53 (2003): 67–71; J. Wainwright, A. J. Parsons, W. H. Schlesinger, and A. D. Abrahams, "Hydrology-Vegetation Interactions in Areas of Discontinuous Flow on a Semi-Arid Bajada, Southern New Mexico," *Journal of Arid Environments* 51 (2002): 319–338.

   The general hydrological phenomena found there have been confirmed by studies elsewhere; see S. J. Schymanski, A. Kleidon, M. Steiglitz, and J. Narula, "Maximum Entropy Production Allows a Simple Representation of Heterogeneity in Semiarid Ecosystems," *Philosophical Transactions of the Royal Society* B 365 (2010) 1449–1455.

15. The ecosystem engineering role of plants is well illustrated by Verboom and Pate's work in eucalypt drylands in Western Australia: W. H. Verboom and J. S. Pate, "Bioengineering of Soil Profiles in Semiarid Ecosystems: The "Phytotarium" Concept. A Review," *Plant and Soil* 289 (2006): 71–102; W. H. Verboom, J. S. Pate, M. A. Abdelfattah, and S. A. Shahid, "Effects of Plants on Soil-Forming Processes: Case Studies from Arid Environments," in *Developments in Soil Classification, Land Use Planning and Policy Im-plications: Innovative Thinking of Soil Inventory for Land Use Planning and Management of Land Resources*, edited by S. A. Shahid (Dordrecht, the Netherlands: Springer, 2013), 329–344.

16. Expansion of tidal channels is illustrated by Z. Zhou et al., "Analysis of the Drainage Density of Experimental and Modelled Tidal Networks," *Earth Surface Dynamics* 2 (2014): 105–116. I have observed high flow chan-nels in various marshes of the US Atlantic and Gulf of Mexico coasts. This paper also suggests mechanisms for developing store-and-pour through

network structures in multidirectional tidal channels: M. Hiatt, E. A. Addink, and M. G. Kleinhans, "Connectivity and Directionality in Estuarine Channel Networks," *Earth Surface Processes and Landforms* 47 (2022): 807–824.

17. J. D. Phillips, "Landscape Change and Climate Attribution, with an Example from Estuarine Marshes," *Geomorphology* 430 (2023): 108,666.

18. For example, see D. R. Belk and J. D. Phillips, "Hydrologic Recovery of Artificially-Drained Wetlands in Coastal North Carolina," in *Coastal Zone '93*, edited by O. T. Magoon (New York: American Society of Civil Engineers, 1993) 3254-3268; B. J. W. Kamrath, M. R. Burchell, J. J. Kurki-Fox, and K. L. Bass, "Impact of Control Structures on Hydrologic Restoration within the Great Dismal Swamp," *Ecological Engineering* 158 (2020): 106024; S. A. Lecce, P. A. Gares, and P. P. Pease, "Drainage Ditches as Sediment Sinks on the Coastal Plain of North Carolina," *Physical Geography* 27 (2006): 447-463; S. A. Lecce, P. P. Pease, P. A. Gares, and J. Y. Wang, "Seasonal Controls on Sediment Delivery in a Small Coastal Plain Watershed, North Carolina, USA," *Geomorphology* 73 (2006): 2460260; R. W. Skaggs, D. M. Amatya, and G. M. Chescheir, "Effects of Drainage for Silviculture on Wetland Hydrology," *Wetlands* 40 (2020): 47–64.

19. This section adapted from this article: J. D. Phillips, "Contingent Partitioning and Adaptation in Hydrological Systems," *Ecohydrology* 16 (2023): e2567. https://doi.org/10.1002/eco.2567.

20. Up to a point, of course. Very large perturbations or changes can rearrange or obliterate any landscape.

21. I will not highlight examples, but you could find this conflation in my own work.

22. For example, see J. D. Phillips, "Thresholds, Mode-Switching and Emergent Equilibrium in Geomorphic Systems," *Earth Surface Processes and Landforms* 39 (2014): 71-79; C. J. Thompson, J. Croke, K. Fryirs, and J. R. Grove, "A Channel Evolution Model for Subtropical Macrochannel Systems," *Catena* 139 (2016): 199-213; K. Fryirs, "River Sensitivity: A Lost Foundation Concept in Fluvial Geomorphology," *Earth Surface Processes and Landforms* 42 (2017): 55-70; S. G. Davidson, P. Hesp, and G. M. da Silva, "Rapid Shoreline Erosion and Dunefield Change, Salmon Hole, South Australia," *Science of the Total Environment* 767 (2021): 145406.

23. Stability analysis methods: Figure 7.12 is a signed, directed, unweighted graph or a signed digraph. A graph interaction matrix A consists of an N X N (N=4 in this case) matrix with entries that are positive, negative, or zero depending on the links between the row and column elements. A has N complex eigenvalues λ_i, the real parts of which are the Lyapunov exponents of the underlying dynamical system; $\lambda_1 \geqslant \lambda_2 \geqslant \ldots \geqslant \lambda_n$. If all $\lambda < 0$ (which must be the case if $\lambda_1 < 0$) the system is stable. If any $\lambda > 0$ (*i.e*, $> \lambda_1 > 0$) the network is unstable.

   $F_k(k = 1, 2, \ldots, N)$ is the feedback at level k of the system, with $F_o = -1$ by definition:

$$F_k = \sum (-1)^{m+1} Z(m, k)$$

$Z(m, k)$ is the product of m disjunct loops with k components. The Fk are equal to the coefficients in the characteristic equation of the system, which can be written

$$F_0 \lambda^n + F \lambda^{n-1} + F_2 \lambda^{n-2} + \ldots + F_{n-1} \lambda + F_n = 0.$$

The system is dynamically stable according to the Routh-Hurwitz criteria if and only if $F_k < 0$ for all k, and successive Hurwitz determinants are positive. The second condition requires $F_1 F_2 + F_3 > 0$, for n=3 or 4. If the first criterion is satisfied, the second indicates that stability is contingent on the feedbacks represented in F_1, F_2 are stronger than those in F_3.

F_i equations are shown below, with signs of the individual a_{ij} not shown. For each scenario analyzed, +. −, or zeroes were assigned to each a_{ij} shown below and the (conditions for) stability determined.

$F_1 = a_{11} + a_{22} + a_{33} + a_{44}$

$F_2 = a_{23} a_{32} + a_{24} a_{24} - a_{11} a_{22} - a_{11} a_{33} - a_{11} a_{44} - a_{22} a_{33} - a_{22} a_{44} - a_{33} a_{44}$

$F_3 = a_{23} a_{34} a_{42} - a_{24} a_{42} a_{33} - a_{23} a_{32} a_{44} + a_{22} a_{33} a_{44}$

$F_4 = -(a_{23} a_{34} a_{42} a_{11}) + a_{23} a_{32} a_{11} a_{44} + a_{24} a_{42} a_{11} a_{33} - a_{11} a_{22} a_{23} a_{33} a_{44}$

24. In twenty-three runoff events in an agricultural watershed, Slattery et al. (2006) documented different modes of runoff generation at the same locations during different events. A Swiss catchment studied by Teuling et al. (2010) is insensitive to precipitation when storage is low, but when storage is high it responds more strongly to inputs. The watershed functions as a simple dynamical system, but more so under wet than dry conditions. With higher inputs, rapid pathways and flow paths are activated, with a more direct link between precipitation inputs and outflows. Rusjan and Mikoš (2015) found that a flysch watershed in Slovenia experienced state changes with respect to flow dynamics according to hydrometeorological conditions. In moist times, the catchment was dominated by dependence on deep storage. But at a threshold rainfall ($\geq$ 10 mm hr-1), secondary streamflow mechanisms were activated, characterized by rapid bypass flow. These are but a few examples of studies showing hydrological system dynamics that switch in different synoptic situations (see reviews by Bonell, 1993; Spence, 2010; and Blöschl, 2022).

M. Bonell, "Progress in the Understanding of Runoff Generation Dynamics in Forests," *Journal of Hydrology* 150 (1993): 217-275; G. Blöschl, "Flood Generation: Process Patterns from the Raindrop to the Ocean," *Hydrology and Earth System Sciences* 26 (2022): 2469–2480; S. Rusjan and M. Mikoš, "A Catchment as a Simple Dynamical System: Characterization by the Streamflow Component Approach," *Journal of Hydrology* 527 (2015): 794-808; M. C. Slattery, P. A. Gares, and J. D. Phillips, "Multiple

Modes of Runoff Generation in a North Carolina Coastal Plain Watershed," *Hydrological Processes* 20 (2006): 2953-2969; C. Spence, "A Paradigm Shift in Hydrology: Storage Thresholds across Scales Influence Catchment Runoff Generation," *Geography Compass* 4 (2010): 819-833; A. J. Teuling, I. Lehner, J. W. Kirchner, and S. I. Seneviratne, "Catchments as Simple Dynamical Systems: Experience from a Swiss Prealpine Catchment," *Water Resources Research* 46 (2010): e2009WR008777. https://doi.org/10.1029/2009WR008777.

25. Some very readable books for the general reader address this. *Finding the Mother Tree* (Alfred A. Knopf, 2021) by Suzanne Simard is written by one of the principal scientists involved in this work. See also *The Hidden Life of Trees* (Greystone, 2016) by Peter Wohlleben and Aliya Whiteley's *The Secret Life of Fungi* (Elliott and Thompson, 2020).

8

Why Everything Is Connected to Everything Else

THE FIRST LAW

It's been called the First Law of Ecology, the First Law of Geography, the First Law of Environmental Science, and it made an appearance in Chapter 1 of this book: *everything is connected to everything else* (EICTEE). The First Law became cited as such and part of public discourse in the early 1970s with the rise of the environmental movement. But the idea goes way back—in all probability to some prehistoric proto-scientist or philosopher. In recorded form, German writer Gotthold Ephraim Lessing wrote in 1769, "*In der Natur ist alles mit allem verbunden; alles durchkreuzt sich, alles wechselt mit allem, alles verändert sich eines in das andere.*" This translates as, "In nature everything is connected, everything is interwoven, everything changes with everything, everything merges from one into another."[1]

Heuristically and pedagogically, the First Law is virtually axiomatic. In real landscapes, it is generally true, though the connections are not always direct, and some can be ignored in some contexts. But when you consider the components of Earth surface systems (ESS) and landscapes and diagram them as a graph or box-and-arrow model, they often come pretty darn close to achieving, if not achieving, full connectivity, where each component is directly linked to all the others by mass or energy fluxes,

Mysterious Ways. Jonathan D. Phillips, Oxford University Press. © Oxford University Press (2025).
DOI: 10.1093/9780197755129.003.0008

cause-effect relationships, genetic and other biological exchanges, spatial adjacency, geographical (e.g., transport or transit) connectivity, and flows and exchanges of information and capital (Figure 8.1).[2]

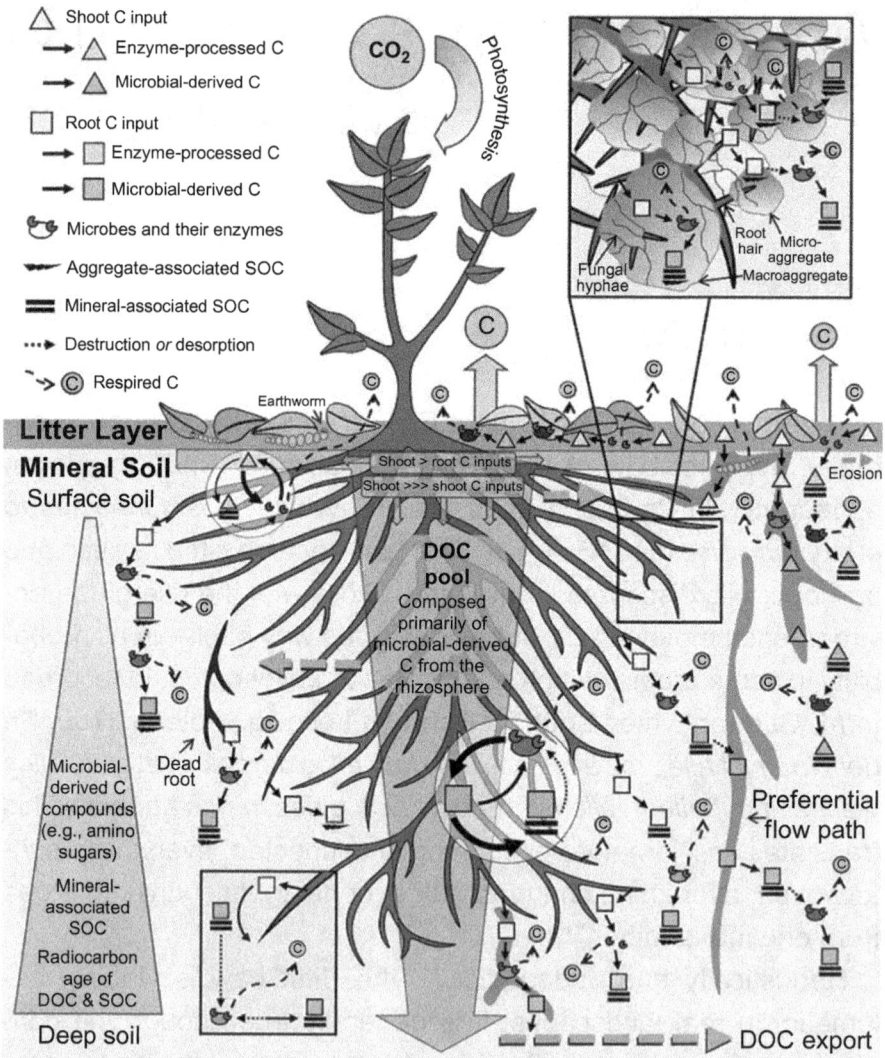

Figure 8.1. The soil carbon cycle. Reproduced under a CC-By-4.0 license from C. D. Gross and R. B. Harrison, "The Case for Digging Deeper: Soil Organic Carbon Storage, Dynamics, and Controls for Our Changing World," Soil Systems 3, no. 2 (2018): 28, Figure 1).

EICTEE in a *direct* sense is typically not possible in spatially explicit representations of landscape systems where the nodes are specific locations or areas and the links are physical flows or exchanges. Spatially explicit examples include graphs of nodes and connecting pathways in transportation networks, stream channel networks represented using discrete channel sources and junctions as nodes and channels as links, and graph representations of the starting and ending points of slope failure and mass wasting events (nodes) connected by the paths of mass movement (links). In the first two examples, it is not possible to travel or flow through the network between some nodes without passing through intermediate nodes. You can't get from Chicago to Los Angeles along Route 66, for example, without passing through all the nodes mentioned in the famous song ("Get Your Kicks on Route 66," composed in 1946 by Bobby Troup and recorded by many). It is not possible to travel along channels from the headwaters to the mouth of a drainage network without going via the nodes in between. And while a landslide could conceivably be connected to multiple source nodes, the sources cannot be directly connected to each other and neither can the endpoints.

So, direct all-connectedness is generally only applicable to *structural* as opposed to spatially explicit representations. Structural graph models are network representations of the structure of landscapes. These may be empirically derived state-and-transition models or correlation structures, or they may be theoretically derived networks based on fundamental equations or system models. The nodes are types or categories of landscape elements connected by flux exchanges, causal influences, or other interactions, or they are system states connected by potential transitions among them (state-and-transition models). A network representation of Figure 8.1, for instance, would be a structural graph. Figure 8.2 shows a more general example.

EICTEE is also potentially applicable to spatial adjacency graphs, a hybrid of spatially explicit and structural types. In these, components (e.g., habitat, soil, or landform types) are connected if they occur contiguously.

Even in structural graphs, connections are often indirect at very broad scales, as many of the types of connections in landscapes

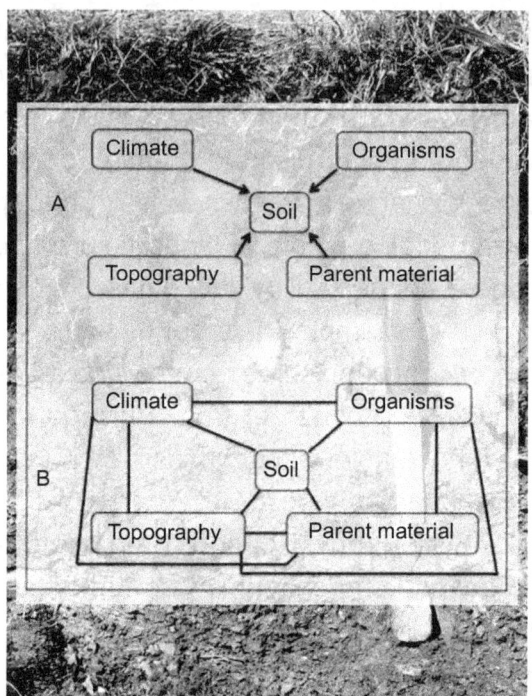

Figure 8.2. (A) is a representation of the state factor model of soils commonly encountered in soil science, ecology, geography, geology, and environmental science textbooks, whereby soils are considered to be a function of parent material (geology), topography, climate, organisms, and time (not shown). (B) is a structural graph showing that soil, parent material, topography, climate, and organisms all affect, and are affected by, each other.

The mutual interactions among state factors and soil were in fact recognized by the fathers of the factorial approach, V. V. Dokuchaev (1846–1903, generally considered the father of pedology) and Hans Jenny (1899–1992). As they were primarily concerned with explaining the variation of soils and soil properties, they expressed the relationships in terms of soils as a function of the other state factors.

depend on spatial proximity. Further, linkages associated with ocean and atmosphere dynamics, global biogeochemical cycles, and trade patterns may be weak at non-local scales. Thus, while logging in Brazil or transportation modes in the United States affect precipitation in Mali and sea ice in the Arctic, connections between a particular forest stand in South America, a farm in North America, a river reach in Asia, and an ice patch in the polar seas are indirect and filtered. Therefore, we are concerned here with the spatial scales (distances, areas, and volumes) and temporal scales of landscapes rather than continental- or global-scale systems.

Some things are connected because they must be—waves and beaches, plants and soil, predators and prey, sedimentary basins and sediment sources, for example. But there exist no laws of nature dictating or promoting system-level connectivity, and it does not have to be so.

High connectivity can indeed be harmful or increase risks in some cases. Global and national economic systems are very highly interconnected, as we were reminded when impacts of the coronavirus pandemic peak in 2020–2021 reverberated around the world and wreaked havoc in nearly all economic sectors. Same for the Russian invasion of Ukraine, still ongoing at this writing. High connectivity of social and political networks and the internet allows rumors, lies, and misinformation to spread at appalling speeds, relatively unchecked. High connectivity sometimes facilitates the spread of viruses (literal or computer), other disease vectors, and pollutants. The high connectivity of Earth's ocean-atmosphere system puts the entire planet at risk (or at least subject to effects) from events occurring in the Arctic Ocean, the Antarctic ice sheets, or the equatorial Pacific. It is just as easy, if not more so, to churn out examples of advantages of connectivity, but it is certainly not an intuitive or logical slam dunk to link connectivity to efficiency or any other benefits to environmental systems across the board. Indeed, the invocation of EICTEE by environmentalists is intended sometimes as an ode to a wonderfully complex nature and its mysterious ways and sometimes as a cautionary warning that degrading or polluting one component or place can have far-reaching negative impacts.

And yet highly connected landscapes develop time after time and place after place. Why? More specifically, why do structural network or graph representations of landscape systems so often develop and maintain very high connectedness? Since nothing external dictates that this be so, what's in it for them?

CROWDED LANDSCAPES

The crowded landscapes concept is based on limits of space and time. Landscapes include a multitude of entities, and many of them are constantly or intermittently in motion. Limited space and time are available for them to exist in, so they inevitably contact each other and become connected in various ways.

Consider any given patch of a terrestrial landscape, say a square meter or so. That patch will be characterized by at least one each of a landform, soil, underlying geology, plant cover, faunal community, microbial community, microclimate, and hydrological status and function. Our patch will be the site of any number of movements and fluxes at, above, and below the surface, and in multiple directions. These will occur with various intensities and frequencies both between and within landscapes and patches. For water alone: gravity-driven flows, capillary rise, root suction, and other biological water use, vapor transport, and state changes. Movements and fluxes also include plant growth (e.g., root extension), faunal digging and burrowing, and critters walking, flying, and crawling around. Some patches will experience transport by wind and mass movement, as well as gravitational settling, and movements due to factors such as shrink-swell and ice formation and melting. Animals (including humans) may deliberately or accidentally transport solid objects or material, and plants may do so locally due to mass displacement.

The point is that even in a single patch of a landscape a lot is going on, along multiple pathways (Figure 8.3). Innumerable opportunities arise for processes and entities to come into proximity and direct contact and to affect each other. Coincidental contacts or influences do not ensure persisting interrelationships, or that everything gets connected, but crowded landscapes do provide necessary, but not sufficient, conditions for the emergence of high connectivity.

Figure 8.3. Vertical and lateral connectivity among surface and subsurface processes, geology, hydrology, soil, vegetation, and other biota (Big Walker Mountain, Virginia). Gravity-driven fluxes occur perpendicular to the ground surface and along and parallel to the ground surface slope. Fluxes also occur along pedological and stratigraphic boundaries, rock joints and fractures, and along and within belowground and aboveground portions of vegetation.

Biological saturation is a common assumption of biogeographic and ecological models. That means that as ecological systems evolve and as new habitats are colonized, something eventually occupies every available niche. An analogous assumption is common in hydrological models and theories—that is, that flow networks become space-filling, expanding to the maximum drainage density that can be supported by the runoff production. Soils and regolith become deeper (though sometimes limited by the depth of landscape incision), though not necessarily thicker, depending on erosion at the surface. Fluvial and karst erosion processes often involve progressive downcutting to their base levels. What these have in common is that all involve growth whereby different elements are more likely to intersect with or encroach upon each other. This in turn indicates that in crowded landscapes (that is, essentially all landscapes) connectivity is likely to increase over time, though this can be delayed, prevented, or reversed by regressive development and clock-resetting disturbance events.

Types of Connections

Table 8.1 shows several different ways that landscape components can be connected. It is undoubtedly incomplete, and some connections could in some instances be varieties or subtypes of others, depending on the scale considered and how much lumping or splitting one is inclined toward. Further, some interrelationships might be placed into multiple categories. So it is with nature, which stubbornly refuses to occupy fixed categories.

Transfers (transportation and transformation) of energy, mass, and information are a common, straightforward connection type. Such connections can be one-way, such as gravity-driven flows or who-eats-who links in food webs. Or they can be two-way, as in tidal fluxes and exchanges of carbon, nutrients, and water in mycorrhizal networks. Transfers may reversible, such as upward or downward heat flux in soil as temperature gradients change, or changes in inputs vs. outputs between stream channels and floodplains at different flow stages, or irreversible due to the role of unidirectional forces. Some transfer connections also involve transformations, as in biogeochemical cycles or state changes of water, while some don't, as in granular or liquid flows. Landscape system components linked by transfer connections can be solely or primarily sources, sinks or destinations, or pass-through, but these roles are sometimes fixed and sometimes not.

Biological interconnections can involve competition, symbiosis (other cooperative-type relationships are lumped in here), and reproduction. Connections between biotic and abiotic elements include mass and energy fluxes but also active and passive ecosystem engineering and abiotic limits on biological, geochemical, and

Table 8.1. TYPES OF CONNECTIONS BETWEEN COMPONENTS IN EARTH SURFACE SYSTEMS

Type of Connection	Examples
Transfer of matter, energy, information	Hydrological flows
One-way vs. two-way	Sediment transport
Irreversible vs. reversible	Nutrient fluxes
Transformative vs. non-transformative	Food webs & trophic pyramids
Source, pass-through, sink	Heat flux
Competition among biota	Competition for resources
	Predator-prey or consumer-producer
	Allelopathy
Biological *reproduction*	Gene flow
	Population dynamics
Partition of mass, energy	Runoff vs. infiltration
	Solar radiation energy budget
	Carbon budget
	Sediment budgets
Mutual adjustments	Hydraulic geometry
	Predator-prey populations
Symbiotic (and other cooperative) relationships	Biological symbiosis & commensalism
Biota	Erosion & weathering
	Fingered flow
Abiotic mutual reinforcement	Weathering fronts

Continued

Table 8.1. Continued

Type of Connection	Examples
Self-limiting relationships (non-biological)	Floodplain elevation vs. overbank flow
	Marsh surface elevation vs. inundation
	Weathering vs. weatherable minerals
Limits	Chemical kinetics
Absolute limits	Base level and fluvial incision
Limiting factors	Population and carrying capacity
Saturation and depletion	Resources: space, water, light, nutrients
Oppositional: fluxes and movements in opposite directions along same pathways	River flows and tides
	Downstream flow and backwater effects
	Reversing winds
	Percolation and water table rise
Process effects	Erosion and vegetation cover
	Evapotranspiration and vegetation cover
	Soil CO_2 and dissolutional weathering
	Slope gradient and gravitational mass flux
Facilitation	Ecological succession
	Biota and preferential flow paths
	Plant-soil interactions
	Passive ecosystem engineering
Self-reinforcement via intermediaries	Active ecosystem engineering

Type of Connection	Examples
Intersections: crossing or convergence of otherwise unrelated movements, fluxes, or features	Faunal trails or pathways Geological joints Antecedent and superimposed streams
Spatial adjacency or contiguity	Soils, landforms, ecological communities, land use
Anthropic Biological, economic, social, political, etc. Deliberate vs. incidental/accidental	See discussion in text

These are not necessarily unique and mutually exclusive. There is not necessarily a direct correspondence between subcategories in the left column and examples in the right.

geophysical processes. Facilitation connections exist where environmental effects allow and promote establishment of subsequent effects. This is well known in ecological succession but also occurs in other contexts, such as unstable wetting fronts and fingered flow phenomena and creation or facilitation of subsurface preferential flow paths by roots and faunal burrowing.

Spatial co-occurrence accounts for other connections, such as spatial adjacency of different landscape elements, intersections of structural features and flow and travel paths, and opposing transport along the same pathways (e.g., reversing wind directions, migration corridors, tidal channels).

The final entry of Table 8.1 lumps together connections caused by humans. Some of these can be accommodated in other categories, such as inputs to agriculture. Others deserve separate consideration due to the strong influence of social, economic, political, and cultural factors. I include them to acknowledge the pervasive effects of human activities on landscapes but do not discuss them in detail to keep the focus on connectivity arising independently of intentional actions or external control.[3]

The landscape components or elements that may be linked by the connections in Table 8.1 include geographical locations or areas; specific landscape features (e.g., landforms, hydrological features, soils, biological communities, geological formations); generalized components of (for instance) hydrological, geomorphological, and ecosystems; individual organisms; taxonomic groups; and process types or regimes.

CONNECTION SELECTION

The crowded landscape concept—limited space, limited time, a tendency toward expansion—explains how elements of landscapes can become connected. But why do some undirected encounters result in functional connections, and why do some connections, and networks thereof, persist and even grow?

The connection selection concept holds that some connections are advantageous. That is, they increase the odds of survival, reproduction, reinvention (see Chapter 7), recurrence, and growth or expansion of at least one of the entities involved. Selection (on average) preserves and enhances connections that increase stability, efficiency, and resistance. It does not do so for those that do not.

For natural selection, this is well established when connectivity arises from heritable traits. At the individual level, if the connectivity trait confers survival or reproductive advantages, it is more likely to be passed on. This can also work at the level of species or higher taxa. Connection selection can also occur as ecological filtering. Habitat characteristics and resources encourage some and discourage other organisms from inhabiting or thriving in a landscape, and they can foster or impede interrelationships. Selection also occurs at the ecosystem level, whereby relationships such as those involved in biogeochemical and energy cycling may be selected for when they maximize efficiency and stability. As we have discussed, selection also operates on abiotic aspects of landscapes.

So, connections are made in crowded landscapes, and connection selection preserves and enhances some links. But (how) is this advantageous to a landscape or network as a whole?

Advantages of Connections

As we already discussed, higher connectivity is not always a good thing. But in landscapes connectivity often is advantageous. If we see the connections as exchanges of information (a definition of information from the *Oxford Dictionary* is "what is conveyed by a particular arrangement or sequence of things"), the state of one component and any changes to it can be communicated to others. This is broadly analogous to the way field indicators can provide information on landscape processes or disturbances.

For biological components, true messages may be involved. Thus, for example, links among components allow a plant to sense an oncoming soil moisture deficit, which may be communicated to other plants via mycorrhizal networks. Responses of ecohydrological systems to environmental changes depend on transfers of information as well as fluxes of energy and matter.[4]

Because many connections involve transfers and transformations of energy and matter, greater connectivity may allow for faster and more efficient transfers by providing more opportunities for partitioning resources. Higher connectivity also provides possible alternative pathways if a transfer connection is closed or degraded. These abilities favor the survival (and possible growth) of the system itself and of individual components.

More links among landscape components also provides more options for adjusting or responding to disturbances and environmental changes. An organism linked to multiple food sources will be better able to adapt to loss of a given source, for instance. The full connectivity among the fundamental hydraulic variables in stream flow allows for multiple modes of adjustment to changes in imposed flows. And a beach linked to multiple sediment sources (e.g., offshore, longshore, and inland such as dune fields) will have more options for recovery from erosion or overwash events. Higher connectivity also promotes synchronization of responses among components, which may be advantageous to the system as a whole. In the stream example above, it allows for near-instantaneous responses to disturbances or changes in inputs.

More links may also confer lower *vulnerability*. If a particular node in a network fails or is lost, more highly connected networks are better able to survive and are not overly dependent on critical nodes whose disruption or loss endangers the entire system. In ecology there exists a large corpus of research on relationships between complexity and stability, which would require a long and complicated discourse to even scratch the surface of. Without doing so, suffice it to say here that the relationship between stability and complexity varies in different ecological systems. But connectivity is only one aspect of complexity, and having a variety of components, no one of which is necessary to the survival of the network/system, reduces the vulnerability to loss, inhibition, or deactivation of any single component. This advantage may be offset by the possibility of bad stuff (e.g., pathogens, pollutants) being more rapidly spread.

A thermodynamics-based principle of evolution grounded on structure, dynamics, and information in complex systems also reveals potential advantages of connectivity. This evolution principle points to configuration of new structures with greater complexity (correlated with but not equal to connectedness) and maximum effectiveness with the minimum elements. This implies evolution toward increased efficiency. Advantages also lie in larger structures, due to the achievement of higher network-wide stability, made of local instabilities.[5]

NETWORKS, GRAPHS, AND EVERYTHING

The components of ESS and the connections between them are readily represented as networks and (mathematical) graphs

(Figure 8.4). Elements or constituents of a landscape system are referred to here as components, corresponding to nodes or vertices in graph theory lingo. Connections between components are links (or edges).

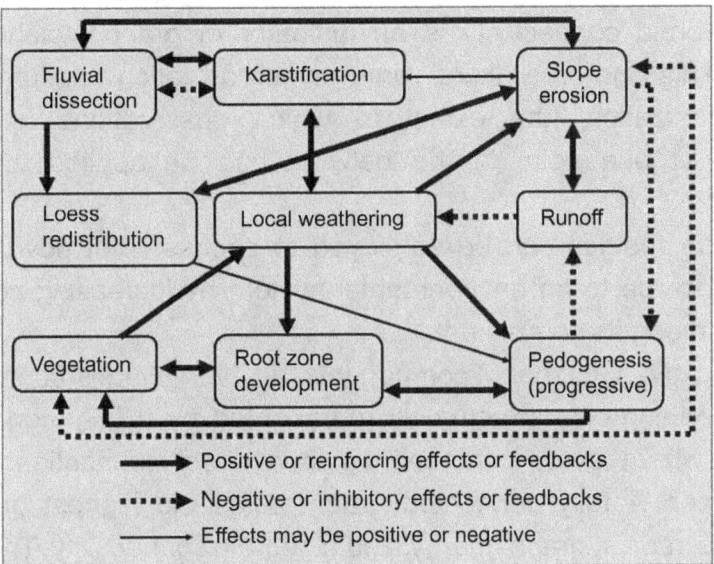

Figure 8.4. Example of interconnections of various types among varied ESS components, based on studies of landscape evolution in fluviokarst areas of central Kentucky, United States. Reproduced with permission from J. D. Phillips, "Why Everything Is Connected to Everything Else," Ecological Complexity 54–55 (2023): 101,051, Figure 1.

A mathematical subfield called algebraic graph theory provides some metrics for evaluating various aspects of graphs and networks. The spectral radius is an indicator of the complexity of the graph. For a graph with a given number of components, spectral radius is sensitive to the number of links among them and to the number of cycles (sequences of links that start and end at the same component, such as the weathering-root zone-vegetation-weathering loop, and the runoff-erosion-pedogenesis-runoff loop in Figure 8.4). Higher spectral radius is associated with greater complexity.

Graph energy, a term derived from applications of graph theory to physical chemistry, is an indicator of the total strength (energy in the physical chemistry context) of positive and negative feedbacks among system components. I often think of it as reverberations or "buzz" of the network. Higher graph energy is associated with networks where information (in the very broad sense mentioned earlier) makes its way rapidly and redundantly through the network.

Algebraic connectivity is an indicator of graph synchronization. This may be literal time synchronization or inferential synchronization (the extent to which observations or inferences at one point in the network can be applied to other portions).

These metrics can be employed to assess what advantages could accrue to an environmental system by increasing connections among its components.

For a given number of components, an everything-is-connected-to-everything-else structure is represented by a fully connected graph where every component has a direct connection to every other. A fully connected graph yields the highest possible spectral radius, graph energy, and algebraic connectivity. This indicates that connect-all configurations have the greatest complexity, the highest degree of total network feedback, and the greatest synchronization.

A fully connected system, compared to other system structures, can respond to change or disturbance with greater rapidity, synchronization, intensity, and with more degrees of freedom. When this is advantageous to network survival, growth, and perhaps propagation, it will be selected for. Higher connectivity also allows negative (with respect to system survival or function) effects to be more widely and rapidly transferred but may also allow more compensatory feedback as well. Rapid transfer in and of itself may be deleterious (as in the spread of contagions or harmful invasive species) but can also be helpful to the system by dispersing or diluting bad stuff such as contaminants. Advantageous connections that arise as ESS components go about their functions are likely to be preserved, moving the network closer to full connectivity.

For the specific type of graph considered in this analysis, the graph energy is directly related to the spectral radius. I derived a method for determining what proportion of a graph spectral radius for a given N (size or number of graph components) is attributable to the number of links, versus the specific arrangement of the links, which I called the wiring. For several ESS I analyzed, the number of connections as opposed to wiring accounts for about 70 percent (65 to 75) of the variability in spectral radius. So, while the increase in the values of the graph metrics is not solely due to greater connectivity alone, it mostly is.

Waccamaw River

But remember the example of a chain standing up end to end—just because the equations say so doesn't mean it is so without some empirical verification. So, let's get out to the real world.

Over a period of about three years, the lower Waccamaw River in South Carolina (Figure 8.5) experienced the three highest flows ever recorded, during major floods in October 2015, October 2016, and September 2018. The 2015 flood was an "atmospheric river" event where moisture from a tropical system well to the south pumped a firehose of moist air into South Carolina, causing extreme rainfall, runoff, and river flooding. The 2016 and 2018 events (Hurricanes Matthew and Florence, respectively) included not only high river discharges but also storm surges from downstream estuaries. Regardless of these large, high-energy flows (and in contrast to the severe impacts on humans and the built environment), geomorphological, hydrological, and ecological changes were minimal. No major state changes occurred in the channel-wetland complex of the lower Waccamaw.

Why not? Because of the high connectivity among hydrogeomorphic components of the system.

I did not directly observe what was happening in the river valley during the floods but have made many field observations since.

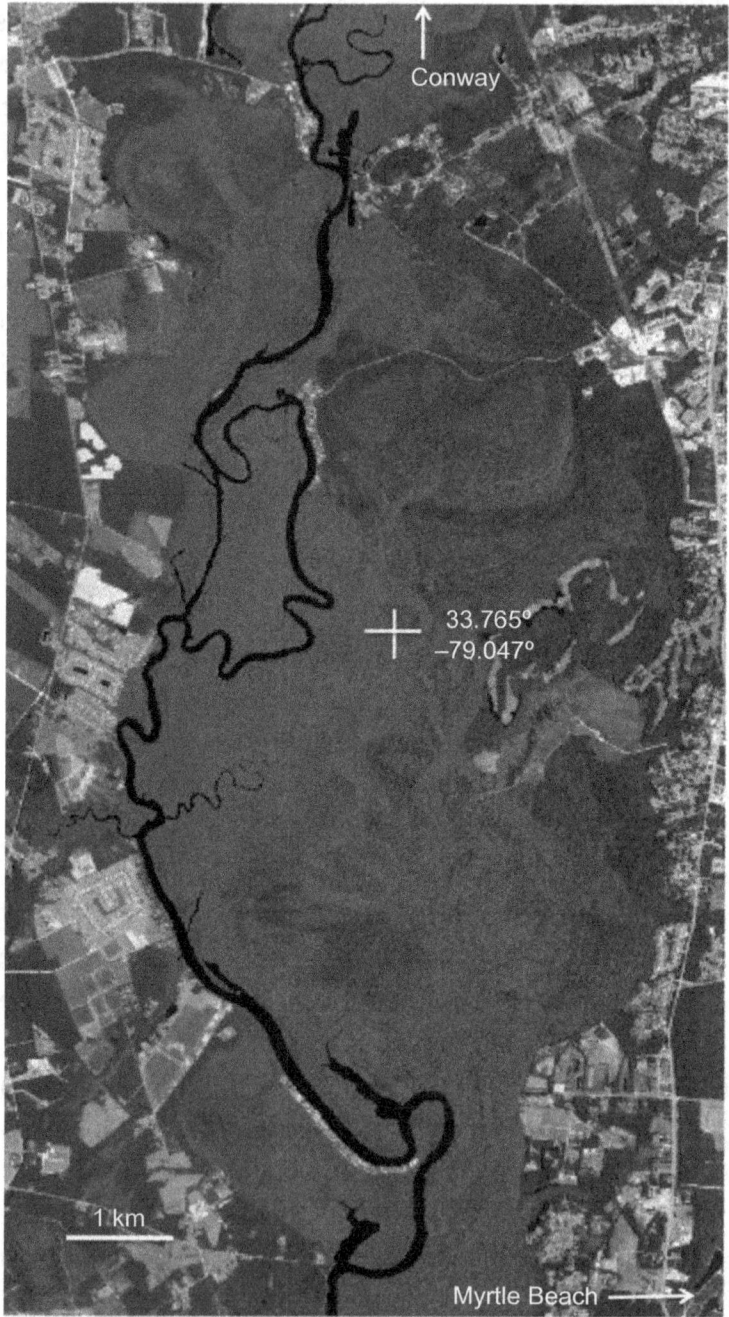

Figure 8.5. Lower Waccamaw River, South Carolina. Only larger hydrogeomorphic features are visible at this scale. Base map: US Geological Survey, National Aerial Imagery Program. Reproduced with permission from J. D. Phillips, "Why Everything Is Connected to Everything Else," Ecological Complexity 54–55 (2023) 101,051, Figure 4.

Further, there is ample aerial imagery—including some during the floods made by the US Geological Survey and the National Oceanic and Atmospheric Administration specifically to assess storm and flood damage. These give a reasonable picture of what was going on at a broad scale, if not a feel for exactly how water flowed through swamps at a specific location.

Based on field observations and imagery, I demarcated the eight types of hydrogeomorphic elements shown in Table 8.2 and patterns of water flow and exchange in Table 8.3.

Flow splits are bifurcations of downstream channel flow; inflows are inputs from external sources (tributaries) or exports coming in from other components. Convergence is when anabranches or distributaries join or rejoin other channels. Backwater effects impede downstream flow or force water upstream. Surface flow across floodplains is cross flow, while return flow is water from floodplains and distributaries going back to channels as stages fall. Local runoff includes surface runoff, groundwater fluxes, and water table rise within the floodplain. Overbank flow is water spilling laterally out of channels, though note that channel banks are generally low, often indistinct, and in some cases are more of a gradual transition from open water to vegetated channel margin to swamp than a distinct bank. Overflow refers to spilling of water from abandoned channel water bodies (ACWBs) to other elements.

Do you see many blank boxes in Table 8.3? There is a high degree of interconnectivity in this system, shown diagrammatically in Figure 8.6. *Almost* everything is connected to almost everything else. The exchanges among elements are in all instances two-way, with the net direction of flux depending mainly on river stages and whether they are rising or falling but also influenced by astronomical tides, local runoff, storm surges, and wind. This enabled the Waccamaw to absorb the flood and storm surge impacts by storing water and delaying flow through wetlands, activating spillways to transport excess water, reversing flows in some components, and to conduct two-way exchanges of water during rising

Table 8.2. HYDROGEOMORPHIC ELEMENTS OF THE LOWER WACCAMAW RIVER

Element	Description
Main channel	Dominant river channel.
Anabranch	Constantly flowing perennial channel roughly parallel to main channel and connected at both ends.
Tributary	Stream flowing into main channel or anabranch. In the study area, all such confluences are within the valley bottom channel-floodplain complex.
Distributary	Streams and channels that normally direct flow from main channel or anabranches to floodplain basins or ACWBs (or further downstream, to estuaries). During falling limbs of floods, flow may temporarily reverse to return flow to channels. Distributaries are perennial; high-flow distributaries are considered flood channels.
Floodplain	Vegetated wetlands (mainly bottomland hardwood swamps) within the valley bottom that are seasonally to frequently flooded. These may convey flow in any direction during high stages, determined by river flows, astronomical tides, local runoff, storm surges, and wind.
Flood channels	Channels that are non-flowing or not inundated during normal or low flows that are activated to convey downstream or distributary flow at high stages.
Backwaters	Channels or ponded areas connected to main channel or anabranches that are normally backwater flooded, with ponded conditions or upstream flow.

Element	Description
ACWBs (abandoned channel water bodies)	Floodplain lakes, oxbows, and sloughs hydraulically isolated from other channels during normal and low flows.

Descriptions refer to the features as they occur within the study area.

and falling river flows (as if you didn't get enough store-and-pour and hydrological partitioning in Chapter 7).

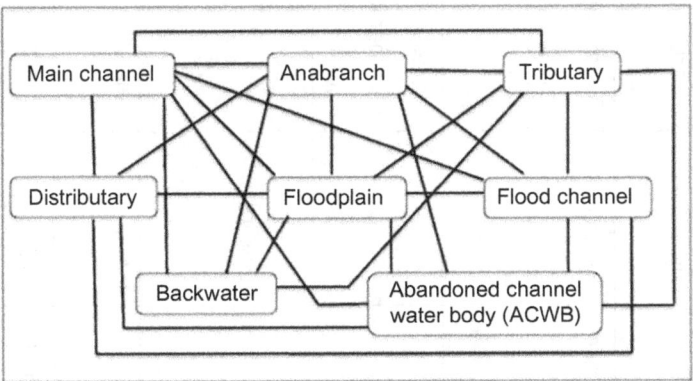

Figure 8.6. Connectivity graph for hydrogeomorphic elements and water exchanges in Table 8.2.

Values of spectral radius, graph energy, and algebraic connectivity are about 90 percent of the maximum possible values for a fully connected $N=8$ graph, indicating the high complexity, connectivity, robustness, and synchronization of the system, and the strong feedbacks within the system. With respect to maintaining the channel-wetland system in the face of variations in water inputs and absorbing the impacts of large floods, the connectivity is highly advantageous. The channel-wetland system has evolved in response to hydrological and geomorphological forcings (such as Holocene and contemporary sea level rise), and the resulting pattern of connectivity is favorable in that context.[6]

Table 8.3. OBSERVED HYDROLOGIC CONNECTIONS AND WATER EXCHANGES AMONG HYDROGEOMORPHIC ELEMENTS OF THE LOWER WACCAMAW RIVER

	MC	Ana	Trib	Dist	FP	FC	BW	ACWB
Main Channel		Flow Split	Backwater	Flow Split	Overbank flow	Flow Split	Backwater	Overbank flow
Anabranch	Flow convergence		Backwater	Flow Split	Overbank flow	Flow Split	Backwater	Overbank flow
Tributary	Inflow	Inflow			Overbank flow	Inflow	Inflow	Overbank flow
Distributary	Return flow	Return flow			Inflow, Overbank flow	Convergence		Inflow
Floodplain	Return flow	Return flow, Cross flow	Return flow, Local runoff	Cross flow, local runoff		Cross flow, local runoff	Return flow, local runoff, cross flow	Cross flow, local runoff
Flood Channel	Convergence	Convergence	Backwater	Inflow	Inflow		Inflow	Inflow
Backwater	Inflow	Inflow	Backwater		Overbank flow	Backwater		
ACWB	Overflow	Overflow	Overflow	Overflow	Overflow	Overflow		

Entries represent fluxes from the row to the column element. See text for explanation of terms.

With respect to an individual ESS component, it is usually advantageous to be directly linked to other components. This facilitates receiving or sending energy, resources, or information and facilitates quicker responses to change. What is the most efficient way for a component to collect information (etc.) from multiple nodes? That would be a convergent radiation pattern where a single key node connected to every other node, which are connected only to the key node; Figure 8.7). Likewise, the most efficient way for a component to distribute or transmit information is a divergent radiation pattern. For an undirected graph, these structures are identical. For any radiation type graph, regardless of N, the spectral radius (often represented as λ_1) = $\sqrt{2} \approx 1.414$ and the graph energy = 2.818.[7] If you want to avoid the numbers and get to the upshot, you can skip down two paragraphs.

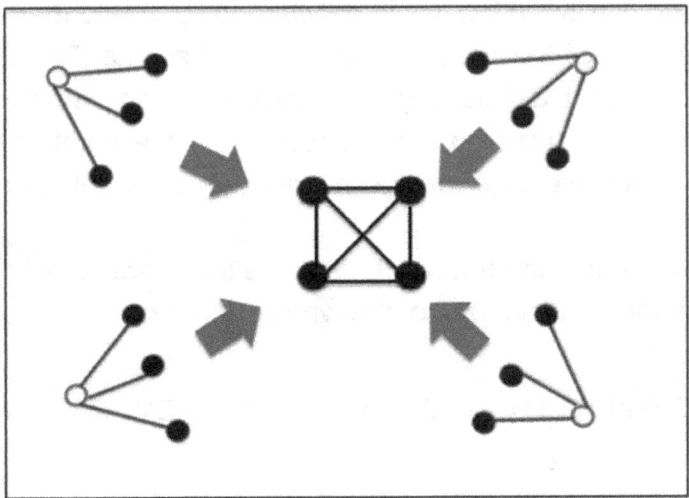

Figure 8.7. Illustration of how four radiation graphs can be combined into a single fully connected graph.

What if we consider a fully connected graph as a collection of radiation subgraphs G_i ($i=1, 2, \ldots, N$) of the fully connected graph G, as illustrated in Figure 8.7? Algebraic graph theory tells us that the spectral radius of a graph is less than the sum of the spectral radii of its subgraphs (symbolically, $\lambda_1 < \sum \lambda_1 G(i)$). If the radiation subgraphs all have N nodes (that is, the same number of components as the parent graph, as in Fig. 8.7), then the sum of the subgraph spectral radii is 1.414 times N (symbolically, $\Lambda = \sum \lambda_1 G(i) = 1.414\ N$).

What all this means is that we can compare the spectral radius for our N components considered as a single fully connected graph $(= N{-}1)$ to that for the same N components as a separate collection of radiation-style subgraphs (1.414 N). For a ten-component system, for instance, the ratio is 0.636, indicating that the spectral radius of the single fully connected graph is less than 64 percent of that of the collected radiation subgraphs.[8]

A fully connected structure is therefore one that *maximizes connectivity* for individual components and for the entire network but that contains subnetworks that *minimize complexity* (as indicated by spectral radius). The fully connected arrangement also minimizes complexity for the whole network compared to the summed spectral radii of the collection of subgraphs. The EICTEE configuration also *maximizes* graph energy for the whole system. Network synchronization also increases as the number of links increases relative to N *and* is maximized for a fully connected network.[9]

Highly connected networks thus serve the interests of individual components or nodes and of the whole network.

WHY *IS* EVERYTHING CONNECTED TO EVERYTHING ELSE?

To put in all in a nutshell, bullet-list style, here's why:

- Connectivity in ESS emerges because hydrological, geomorphological, pedological, biological, and climatological processes operate continuously in landscapes, transporting and transforming mass, energy, and information. Their zones of operation inevitably encroach on each other, and the transport pathways intersect. This creates connections among the elements or components involved. This is the *crowded landscapes* idea.

- Landscapes often get even more crowded because flow networks often evolve to get longer and denser, ecological niches to get fully occupied, and soils to get thicker. This makes landscapes even more crowded as various, often-growing, elements come into contact.
- Connections may get preserved and reinforced due to positive feedbacks (e.g., enlargement and maintenance of preferential flow paths), mutually beneficial effects on interacting components (e.g., soil-plant interactions), the irreversible nature of some processes and pathways (e.g., decomposition, weathering, slope movements), and efficiency advantages of using already-established pathways. This is *connection selection* at the level of individual links.
- Connectivity between elements often has net benefits to the individuals. This can lead to connection selection at the component (or node) level.
- Increased connectivity of the landscape system has advantages of more intense and rapid energy and material cycling, more degrees of freedom for adapting to change, and greater synchronization of evolution and responses. Disadvantages of increased complexity are (usually) offset by these benefits and the simultaneous achievement of maximum network efficiency for individual components and reduced cumulative complexity for the whole system. This is connection selection at the network level.

Organisms can't exist apart from their ecosystems, so connectivity within biological communities and between biota and abiotic environments is well established. The tight connectivity of biota has become even clearer in recent years as researchers have found, for example, that plants and microbes of both the same and different species can communicate directly or via biological intermediaries—for example, the "wood wide web" that connects forest organisms. Biotic interactions (i.e., connections) have been identified as an important component of biodiversity. A group from Mexico's Instituto de Ecología developed a framework for studying interactions and biodiversity vs. ecosystem function relationships by identifying the unique and common interactions, highlighting a key role for connectedness in ecosystem functions.[10]

For abiotic components of ecosystems and biotic-abiotic inter-actions, advantages of faster biogeochemical cycling were rec-ognized as least as far back as Lotka in 1922. Two threads of evidence support this. For individuals, more rapid use of water, nu-trients, and food is associated with, and supports, faster growth and general health and thus favors survival and reproduction. Second, analogous to the economic advantages of keeping capital in circu-lation, the total availability of a fixed amount of matter and energy is increased the more and faster it is cycled among ecosystem components.

Evolution of highly connected hydrological flow networks pro-duces store-and-pour configurations (Chapter 7), as exemplified by fluvial channel networks, preferential flow paths in soil, and dual-porosity groundwater systems. These configurations enhance the stability of both surface and subsurface hydrological systems by allowing them to better handle both low- and high-input (dry and wet) episodes, as illustrated in the Waccamaw River example above. Ecohydrological feedbacks, especially via plants, provide positive feedback to store-and-pour structures. More rapid use and cycling of water has also been shown to confer ecological advantages.[11]

Studies of plant-soil reciprocal interactions (vs. edaphic prop-erties of soils with respect to plant growth) have highlighted the mutual interactions of plants and soils (including the microbial, fungal, and faunal components of soil) in the context of ecosys-tem engineering and niche construction. But those interactions also benefit soils by enhancing resistance and resilience via, for example, erosion protection, aggregate formation, and store-and-pour soil hydrology.

While the key role of interconnections involving geological and geomorphological and geological components of ESS is quite clear, it is admittedly not intuitively evident how these benefit the landforms and geological features. Indeed, weathering breakdown of parent rock is a prerequisite for the formation of regolith and soil. Erosion, sediment transport, and deposition may have positive effects on ESS, but often these are negative for preservation of both geomorphic and other components. However, it is possible that the connections with biota, soil, and hydrology both enhance the preservation of landforms and are necessary for geomorphic recovery from disturbances.

If we accept that connectivity is usually advantageous for ESS networks, (how) does selection preserve and enhance connections? In geomorphological and hydrological processes this often occurs due to gradient, resistance, and efficiency selection. Pathways with the steepest flux gradients, landscape elements with the highest resistance (to, for example, weathering and erosion), and configurations with the greatest efficiency for work (consistent with the least action principle) preferentially occur, recur, grow, and survive as described in Chapter 6.

In ecosystems, biogeochemical cycling, and abiotic-biotic interconnections, Darwinian natural selection selects for connections that advantage individual biota. Because organisms cannot survive independently of their ecosystems, the ecosystem is the primary unit of selection, some ecologists and biogeographers have argued quite persuasively.[12] I have previously argued that ecosystems respond to climate change as integrated units, not as a collection of individual elements, which increases the odds of preservation of ecosystems.[13]

Perhaps everything is connected because it just is or because some guiding force wanted it that way. But there are net benefits in high connectivity. If multiple individual components connected to others to maximize communication, information, energy, and mass fluxes, and so on—to the benefit of those individuals—are brought together in a maximum-connectivity structure, it is also beneficial to the whole landscape system. Selection would thus tend to favor such arrangements, allowing for the occasional case where high connectivity backfires and for the nondeterministic nature of selection. Everything is connected to everything else (more or less) because high connectivity works—and things that work tend to proliferate and persist.

NOTES

1. Leonardo da Vinci (1452—1519) is supposed to have written that everything in nature is interconnected, but the evidence is sketchy. The Lessing quote comes from *Hamburgische Dramaturgie*. Biologist Barry Commoner (1917–2012) used the EICTEE phrase in print as early as 1969, dubbed it the First Law of Ecology in 1970, and included it as such in his 1971 book *The Closing Circle: Nature, Man and Technology*. EICTEE was cited as the First Law of Geography in W. R. Tobler, "A Computer Movie Simulating Urban Growth in the Detroit Region," *Economic Geography* 46 (1970): 234-240. I first encountered it as a college student in the 1970s, via K. E. F. Watt's *Principles of Environmental Science* (1973).
2. Much of the material in this chapter is based on this article: J. D. Phillips, "Why Everything Is Connected to Everything Else," *Ecological Complexity* 54–55 (2023): 101051.
3. And, frankly, to avoid the extensive and often exhausting parsing of terminology that often occurs in social sciences and humanities.
4. Communication in mycorrhizal networks is covered in Suzanne Simard's book *Finding the Mother Tree: Discovering the Wisdom of the Forest* (New York: Knopf, 2021). The ecohydrology principle is based on A. E. Goodwell, P. Kumar, A. W. Fellows, and G. N. Flerchinger, "Dynamic process connectivity Explains Ecohydrologic Responses to Rainfall Pulses and Drought," *Proceedings of the National Academy of Sciences (USA)* 115 (2018): https://doi.org/10.1073/pnas.1800236115.
5. J. L. Usó Doménech et al., "Structure, Thermodynamics and Information in Complex Systems," *Kybernetes* 51 (2022): https://doi.org/10.1108/K-09-2021-0858, https://www.emerald.com/insight/0368-492X.htm.
6. Phillips (2022), endnote 2, also gives an example of a poorly connected system that was not maintained and was transformed by Hurricane Florence.
7. Spectral radius is the largest eigenvalue (λ_1) of the interaction matrix of the system graph. Graph energy $E(g)$ is the sum of the absolute values of the eigenvalues of the adjacency matrix.
8. That is, $\lambda_1/L = (N - 1)/1.414N$. For $N = 10, 9/14.4 = 0.636$.

9. Technical/mathematical note: synchronization is measured by algebraic connectivity a, which is equal to the second smallest eigenvalue (largest non-zero eigenvalue) of the Laplacian of the graph adjacency matrix:

$$a = \lambda(L)_{N-1}$$

The Laplacian is

$$\boldsymbol{L} = \boldsymbol{D} - \boldsymbol{A}$$

where $\boldsymbol{D}$ is the degree matrix of the interaction matrix $\boldsymbol{A}$. Algebraic connectivity is constrained by

$$4/ND \leq a \leq \kappa(G)$$

D is graph diameter (the minimum longest path between any pair of nodes) and $\kappa(G)$ is vertex connectivity, the minimum number of nodes that could be removed to disconnect the graph. Maximum possible algebraic connectivity is associated with a fully connected graph, where $a = N-1$. Though the addition of a single link might leave graph diameter D unchanged, it cannot reduce it, and continued adding of links must lower D by reducing the maximum shortest path between any two nodes, thereby increasing the lower bound of a. At the same time, more edges or links must also increase the vertex connectivity.

10. P. Luna, E. J. Corro, R. Antoniazzi, and W. Dattilo, "Measuring and Linking the Missing Part of Biodiversity and Ecosystem Function: The Diversity of Biotic Interactions," *Diversity* 12 (2020) 86; https://doi.org/10.3390/d12030086.

11. For instance, see Z. Cong et al., "Ecohydrological Optimality in the Northeast China Transect," *Hydrology and Earth System Sciences* 21 (2017): 2449-2462; M. del Jesus, R. Foti, A. Rinaldo, and I. Rodriguez-Iturbe, "Maximum Entropy Production, Carbon Assimilation, and the Spatial Organization of Vegetation in River Basins," *Proceedings of the National Academy of Sciences (USA)* 109 (2012): 20,837–20,841; P. S. Eagleson, *Ecohydrology: Darwinian Expression of Vegetation Form and Function* (New York: Cambridge University Press, 2022); Y. Zhang, T. Zhao, C. Shi, and Q. Ma, "Simulation of Vegetation Cover Based on the Theory of Ecohydrological Optimality in the Yongding River Watershed, China," *Forests* 12 (2021): 1377. Also see endnotes 7 and 8 from Chapter 7.

12. This idea was discussed in Chapter 2.

13. J. D. Phillips, "State Factor Analysis of Ecosystem Response to Climate Change," *Ecological Complexity* 40(A) (2019): 100789.

9

Perfection

Science fiction writer Arthur C. Clarke famously wrote that "any sufficiently advanced technology is indistinguishable from magic." I was riffing on that in a talk I once gave where I maintained that "any sufficiently improbable event is indistinguishable from a miracle." The presentation was titled "Geomorphic Miracles in the Arms of God"—the sort of quirky title I love but that also causes a certain amount of eye-rolling. The arms of God bit referred to my study area: the Brazos River, Texas. It was originally named by Spanish sailors who, spying its muddy freshwater plume while being battered by a storm in the Gulf of Mexico, found salvation by sailing up the river. They named it *Brazos de Dios*, or arms of God. In the geomorphic miracles bit, I was making the case, using examples from the Brazos, that almost any geomorphic landscape is the result of multiple interacting environmental controls and a path-dependent contingent history. The joint probability of the same combination recurring anywhere else or at any other time is vanishingly small—so low that it is tantamount to miraculous.

My thinking along these lines was also influenced by another sci-fi author, Alan Moore, in the graphic novel *The Watchmen*: "In each human coupling, a thousand million sperm vie for a single egg. Multiply those odds by countless generations, against the odds of your ancestors being alive, meeting, siring this precise son; that exact daughter . . . until your mother loves a man . . .

Mysterious Ways. Jonathan D. Phillips, Oxford University Press. © Oxford University Press (2025).
DOI: 10.1093/9780197755129.003.0009

and of that union, of the thousand million children competing for fertilization, it was you, only you . . . (it's) like turning air to gold . . . a thermodynamic miracle." I read *Watchmen* as it came out originally in twelves issues in 1986–1987 (just establishing some comic book and science fiction nerd street cred here), when my scientific ideas on contingency were no more than embryonic. However, as those ideas developed, the thermodynamic miracle passage stuck with me.

Another metaphor helped me tie it all together.

THE PERFECT STORM

A mid-level low-pressure system developed over the Atlantic Ocean south of Bermuda on October 23, 1991. The developing storm was initially designated a subtropical storm by the US National Weather Service (NWS) on October 26. Meanwhile, a mass of clouds near Bermuda grew steadily more convective and was swallowed up by the low-pressure circulation. The system gained tropical cyclone status on October 27. NWS named it Tropical Storm Grace, which soon grew into a hurricane, Category 2 on the Saffir-Simpson scale.

On October 28, low pressure developed off the coast of Nova Scotia after a strong cold front plowed through. While the low rapidly intensified the next day, a strong high-pressure area was building over eastern Canada. The proximity of the strong high- and low-pressure systems resulted in steep pressure gradients. The extratropical low began to suck in Hurricane Grace on October 29, directing it just south of Bermuda, absorbing its tropical moisture, and entraining Grace in the eastern side of the huge extratropical circulation. Because of its origin in the rare, highly improbable concurrence of a strong high and two strong low-pressure systems, one of which itself had an unusual genesis as a tropical cyclone with a subtropical origin, NWS dubbed it the "perfect storm."

The storm peaked in intensity on October 30 and again exhibited another atypical trait for a tropical cyclone, drifting southwest
toward the New England coast. From Canada down to North Carolina, waves up to 10 m high were observed. Extensive storm surge
and wave damage occurred along the east coast, with water levels
reaching their highest points in half a century at some locations.
The 21 m fishing vessel *Andrea Gail* was among the numerous
craft sunk by the perfect storm. The *Andrea Gail* went missing on
October 28; neither the six crew nor the ship were ever found. Se-
bastian Junger (1997) wrote a best-selling book about the ship,
the subsequent search, and likely causes of its demise. A film version of the book was released in 2000; both were titled *The Perfect
Storm*.

But the weirdly perfect storm was not done. A sub-circulation
within it strengthened into a full-on hurricane—never named because NWS did not want to alarm and confuse the public in the
wake of the recent storm damage. The Hurricane Whose Name
Shall Not Be Spoken remained at sea and made landfall as a weak-
ening tropical storm in Nova Scotia. Even as the Perfect Storm was
spinning off a nameless hurricane, the upper Midwest of the United
States endured a massive blizzard. The storm system formed over
the far western Gulf of Mexico, strengthened, and moved north-
northeast toward Minnesota and Wisconsin on October 31 and
November 1. The track was unusual (as opposed to the typical
west to east) because the Perfect Storm was serving as an atmospheric block to the east. The Halloween Blizzard, as it came to
be called, was packed with Gulf of Mexico moisture and produced
record snowfalls, while 50 to 80 km hr^{-1} winds created blizzard
conditions.

After the storm(s) of late October 1991, and even more so after
Junger's book and the subsequent film, *perfect storm* came into
general use as a metaphor for an unusual, improbable convergence of circumstances that produces large and singular results.
Popular use of the metaphor usually signifies negative, if not catastrophic, outcomes for humans. It dawned on me that perfection in
this sense of extreme improbability (thermodynamic miracles?) is

an apt description of nature. The perfect storm of 1991 was partly governed by fundamental, invariant laws of atmospheric physics, and Earth surface systems (ESS) are likewise governed by general laws independent of time and place. The laws operate, however, within a context of resources and boundary conditions that are particular, if not unique, to specific places. Further, those law-place interactions are conditioned by history. The applicable laws themselves are not variable, but which set of them are germane, and their relative importance, varies from place to place and time to time. The place factors are distinct—unique at some level of detail. And the historical circumstances are unique—think of chains of episodes and events and disturbances. Thus, *every* landscape represents an unusual, highly improbable outcome, at least in terms of being duplicated at other times and places (Figure 9.1).

I had been analyzing geographical and historical contingency, and their relationship to global laws, for a while. I published my first explicit discussion/analysis in 2001, a piece titled "Human Impacts on the Environment and the Primacy of Place." The next year I produced a more rigorous analysis of the relative contributions of global (laws) and local factors on ESS. At that stage, I lumped place and history together in the "local" category. This lumping persisted in print until 2007, when I adopted the "perfect" metaphor for ESS and landscapes. In 2007, "The Perfect Landscape" appeared in *Geomorphology*, with due credit to the perfect storm metaphor, and separation rather than lumping of geographical/place and history factors.[1]

Box 9.1

THE PERFECT LANDSCAPE FORMALISM

The perfection concept is probabilistic, where *p(S)* is the probability of the existence of a given landscape:

$$p(S) = \prod_{i=1}^{n} p(X_g) \prod_{j=1}^{m} p(X_p) \prod_{k=1}^{q} p(X_h)$$

X_g, X_p, X_h, respectively represent the $i = 1, 2, \ldots, n$ applicable global or general laws, the $j = 1, 2, \ldots, m$ relevant place factors, and the $k = 1, 2, \ldots, q$ historical contingencies (history factors). The $p(X)$ indicate the probability of occurrence of each law, place, and history factor—i.e., the likelihood of occurrence in a given landscape. Many $p(X_g) = 1$, but by definition all $p(X_p) < 1$, and many $<< 1$. Most $p(X_h) << 1$. Multiplying numbers <1 means $p(S)$ gets very small, very quickly. Furthermore, the greater the number of factors or variables included, especially place and history factors, the smaller $p(S)$ gets. Including more

Figure 9.1. Artist Petr Mores, in consultation with pedologist, ecologist, and geomorphologist Pavel Šamonil, produced this figure to illustrate tree-rock-soil-landform interactions in old-growth forests of central Europe. Unique combinations of underlying geology, tree species, soils, landforms, microclimate, and histories of climate change, land use, and disturbances produce plant-soil and biogeomorphological interactions that have some similarities in other forests but are unique to any given forest landscape. The elements in this figure are based on Boubinsky Prales, Czech Republic.

> factors, even laws, can only reduce and cannot increase $p(S)$. Thus, the more detail you incorporate, the more improbable the landscape becomes, indicating uniqueness and perfection.

As far as I know, the perfection formalism (see Box 9.1) and terminology are original. The basic idea, not so much. Observers of nature have long, probably always, known that no two Earth system entities are identical in detail. Many scientists recognized that, without having to drill all the way down to minute details, irreducible elements of individuality exist in environmental systems. This may enchant or confound (depending on one's point of view) science's collective effort to explain as much as possible—ideally everything—with global laws.

In 1963, George Gaylord Simpson (1902–1984) distinguished *configurational* and *immanent* factors in geology. The latter are general and ubiquitous. Configurational elements, by contrast, are historically contingent factors emanating from interactions of immanent factors (laws operating in geographical/geological contexts) with historical circumstances. Simpson argued that geological events are unique (perfect, I would say) because immanent/law factors operate within idiosyncratic historical and environmental contexts.

Stanley Schumm (1927–2011), one of the great thinkers of twentieth-century geology, considered place and history factors as *singularities*, the characteristics that make each geological entity unique to some extent. In his 1991 book *To Interpret the Earth: Ten Ways to Be Wrong*, Schumm argued that singularities are a fundamental trait in Earth science. He maintained that responses of geomorphic systems to changes or disturbances must always vary to some degree even between apparently similar systems because of these singularities.

Soil individuality—at the pit or pedon level—has long been recognized, even (or maybe especially) by those trying to tease out or discover generalities and similarities in soil geography and geomorphology. Charles Kellogg (1902–1980), longtime chief of soil survey operations for the US Department of Agriculture, wrote in 1956 that "each soil has its own history. Like a river, a mountain, a forest, or any natural thing, its present condition is due to the influences of many things and events in the past." In 1967, Raymond Daniels (1925–2009), a prominent figure in pedology and soil geomorphology, reported in a scientific article the conventional wisdom of soil surveyors in the US coastal plain region that "if you want to show someone the soil you saw while mapping, you must return to within six inches of your auger boring."

TRIANGLES, BADASSES, AND AXIOMS

Landscape perfection is not inconsistent with the continued search for laws and generalizations—laws are a key part of perfect landscapes—but perfection does emphasize the inescapable role of geography and history and the need to account for chance, history, and locality in understanding landscape evolution. Beyond recognizing irreducible contingency, what does perfection imply for the way we study, conceptualize, and comprehend landscapes? In trying to address this, I have proposed three overlapping approaches: the laws-place-history triangle, badass geomorphology, and axioms for landscape interpretation. The triangle was outlined in Chapter 1.

Badass Landscapes

My scientific colleagues tend to be either appalled or amused by this terminology. Though badass may be vulgar and colloquial, the badass archetype or trope frequently describes, even glorifies, deeds of individualistic anti-authority characters who achieve disproportionate changes or results. Metaphorically, these are indeed the traits of perfect landscapes—rebellious in that they frequently defy conventional explanations, individualistic by definition, and frequently featuring dynamical instabilities that produce disproportionate changes. Using this rationalization, an article titled *Badass Geomorphology* appeared in 2015.[2]

In 1939, H. A. Gleason (1882–1975) developed an individualistic concept of plant associations that was a key theoretical development in ecology. Drawing on Gleason's concept, I proposed a broadly analogous individualistic concept of landscape evolution based on three necessary conditions. First, the concept applies when and where there exists positive evolution space— that is, mass, energy, space, and time sufficient for development to occur. The second condition is axiomatic: every landscape component can change or evolve. The third condition is that the environment varies, exerting selection pressure. The third condition means that as conditions change, some features, forms, processes, and patterns are comparatively enabled or stimulated while others are inhibited. This includes, and is exemplified by, the ecological and geomorphological filtering processes described earlier.

The individualistic concept states that there is "room" for change, that landscape components are changeable, and that variable environments select for different landforms. While this doesn't dictate that any pair of landscapes must differ, it does mean that similarity would require comparable evolution spaces and common environmental selection pressures. Where two systems are similar, it is interpreted as a happenstance of similar environmental selection rather than a predetermined stage or climax of landscape evolution, a normative equilibrium, or other attractor state. Ecologists will recognize this as directly analogous to Gleason's framework, which maintains that occurrence of similar plant communities is a circumstantial result of similar environmental pressures operating on similar sets of species populations.

My 2015 paper used the "badass bends" of the Kentucky River gorge as a real-world example of an explicit application of the badass framework. New Zealanders Ian Fuller (Massey University) and Michael Marden (Landcare Research) and colleagues applied the badass concept and terminology to gullies on the North Island of New Zealand.[3]

One could just as easily and justifiably, of course, refer to badass ecology, badass hydrology, badass pedology, and so on.

Axioms

Understanding ESS ultimately comes down to reading the landscape, which has been referred to as a "dark art."[4] There is more than one way to read a landscape. We can read it as history, teasing out as best we can how the landscape came to be and the events and changes along the way. We can read it via laws, in terms of how those laws are imprinted on it and embodied within it. We can read the landscape from a place perspective, disentangling the multitude of environmental and geographic properties and contexts that influence ESS and then trying to weave them back together into a coherent story of how and why the landscape is the way it is. These readings are rarely fully independent of each other, but the perfect landscape perspective and the law-place-history (LPH) framework show that reading the landscape should involve all three approaches.

But how to transform this from rhetoric to research, from pedantics to practice, from dark art to science? Clearly it can be done, because many have accomplished it, but as far as I could tell the how-to was tacit knowledge. So, in a 2018 article I tried to combine perfection-related concepts and distill from them from them some principles or axioms for reading landscapes. The key concepts are mainly ideas discussed in this book: multiple causality and polygenesis, the LPH triad, individualism (or badassery), evolution space, and selection. To this I added a concept borrowed from human geographer Allan Pred (1936–2007).[5]

"Place as Historically Contingent Process," is the title of a 1984 article by Pred, who was concerned with social science issues such as divisions of labor and power relations. Details of his conceptual model are not applicable to biophysical landscapes. However, place as historically contingent process (PAHCP) is applicable to place formation (landscapes are places, after all), and historical contingency is explicit. Studies of landscapes typically depict change as historical sequences or cycles, or state-and-transition models, while Pred conceptualizes place-making in terms of multiple processes, controls, and influences acting simultaneously or contemporaneously and merging or melding into one another. That is a pretty good description of landscape evolution.

PAHCP is based on four proposals. First is constant "becoming," where places are viewed as plastic and malleable, and the observed condition is seen as a sample of a constantly changing entity rather than a static state—a still photo from a never-ending film. Second, processes on one hand and structures on the other are intricately interwoven, such that they constantly "become one another." This reflects the mutual adjustments among processes and structures or forms (think of food webs, processes in and morphology of soil profiles, or hydraulic geometry). Third, environments, historical imprinting, and transformations are also "becoming one another," as changes become historical legacies that become part of contemporary environmental contexts. Finally, the dynamics of process-structure and environment-history-transformation occur simultaneously, not sequentially.

The self-evident parts of the key concepts can be presented as ten axioms, shown in Table 9.1.

I applied and illustrated the axioms in a study of evolution of soil landscapes of central Kentucky. No one else has explicitly applied the axioms and admitted it in print, but some published work can be readily mapped onto the axiomatic framework. One example

Table 9.1. AXIOMS FOR LANDSCAPE INTERPRETATION

---

1. *Spatial structuring happens*, and spatial differentiation occurs.

2. *Selection occurs* as some structures and patterns are preferentially preserved and enhanced.

3. *Coalescence*: structuring and selection weave together sections of space into zones (places). These places are internally defined or held together by mass or energy fluxes or other functional relationships and/or characterized by distinctive internal similarity of traits.

4. *Individuality and constraints*: places have unique, individualistic (perfect) aspects; however, development is bounded by an evolution space defined by applicable laws and available energy, matter, and space resources.

5. *Mutual adjustments*: mutually adjusting (reciprocal) interrelationships exist between process and form (pattern, structure), and among environmental archetypes, historical imprinting, and environmental transformations. These operate at various rates but occur constantly and contemporaneously.

6. *Canalization*: place formation is increasingly constrained between clock-resetting events.

7. *Constant change*: stable, even static, states can be observed over certain timescales and periods, but places are always changing or subject to change.

8. *Reversibility*: spatial structuring, divergence, and place-making are reversible. Places can merge or coalesce or be obliterated by CREs or convergent-divergent mode shifts.

9. *Scale range*: all phenomena above may occur, or be observed or analyzed, across a range of spatial and temporal scales.

---

is Marden and Fullier's studies of badass gullies in New Zealand, which are based on the same conceptual frameworks underpinning the axioms. Johnson and Ouimet's (2018) method for interpreting landscapes through airborne light detection and ranging (LiDAR) is another. Their explicit concern with landscapes as palimpsests makes it clear their interpretive approach is consistent with the axioms. An analysis of barrier island state spaces by Li-Chih Hsu and J. A. (Tony) Stallins also makes use of the axiomatic framework.[6]

Another study consistent with the axioms concerns Holocene evolution of tidal systems in the Netherlands by a group from Utrecht University. Land subsidence, tidal fluxes, river water and sediment inputs, avulsions, and flooding by rising sea level are examples of spatial differentiation and structuring operating on a partially inherited landscape. Selection differentially enhances or preserves (or removes or obscures) some features, and coalescence is observed in the form of, for example, estuaries, lagoons, marshes, flats, dunes, cover sands, and river floodplains. Individuality asserts itself via glacial impacts, inherited topography, isostatic rebound, human influences, and offshore sediments. But evolution space defined by tidal range, base level, and laws of sediment transport, and deposition constrains the individuality. Important mutual adjustments in the Dutch tidal systems include vegetation-sedimentation interactions and sedimentary infilling rates versus tidal prism feedback. Constant Holocene relative sea level rise partly canalizes the evolution. The axioms of reversibility, scale dependence, and constant change are evident throughout the dynamic Holocene development.[7]

EVOLUTIONARY CREATIVITY

Perfection implies creativity (Box 9.2). Perfect landscapes are unique in non-trivial ways, and their existence implies production

of novel geomorphic, ecological, and soil systems in addition to reproducing existing ones. Novelty is plainly evident in biological evolution, and organisms certainly influence abiotic aspects of landscapes. But can novelty emerge independently of biota? Creativity
in biological evolution is evident as the possibility for selection
and evolutionary dynamics to produce new entities. In organisms,
this is mainly due to genetic processes—mutation and genetic
drift—which are absent in abiotic phenomena. With no genetic
memory, the repeated reinvention of adaptive mechanisms such as
store-and-pour by abiotic systems represents creativity, but while
there is room for variation, reinvention does not necessarily involve
novelty.

Box 9.2

THE LEXICON OF EVOLUTIONARY CREATIVITY

Words such as *creativity*, *innovation*, and *adaptation* have
broad, multiple meanings in different contexts. Here's how I use
these and related terms here.

Adaptation, defined generally as the action or process of
adapting or being adapted, in evolutionary science is usually
associated with "the process of change by which an organism or species becomes better adapted to its environment." In
a broader context, adaptation is adjustment of environmental
systems to enable survival and efficient functioning.

Adjustment is a synonym of adaptation, and as a term that
is generally more acceptable in a geoscience context it is used
here.

Creativity in dictionary definitions is mainly related to deliberate human efforts, though the most general definition of *creation*
is simply "the action or process of bringing something into existence or prominence." In complexity research (for instance,
studies of how complexity arises in living systems), creativity is
used more broadly to refer to the appearance of new entities

or phenomena. I use it in that sense here. Creativity in landscapes is thus defined here as the capacity for emergence of new entities—specifically those that increase the adjustment of the landscape to environmental conditions.

Emergence is defined as the process of coming into existence or prominence by the *Oxford English Dictionary*. This is a suitable term for the appearance of new entities, though it should not be conflated with more specific concept of emergence from the internal dynamics of complex systems. The latter applies to some, but not all, emergence in landscapes. Emergence does not distinguish whether the new phenomena are improvements or adaptations.

Novelty is the appearance or development of new forms or features.

Conscious, purposeful creation is impossible for abiotic components of landscapes. For that matter, in most organisms adaptations and improvements in evolutionary fitness arise from inheritable traits, not purposeful actions. Emergent entities (forms, structures, patterns, processes) are, however, subject to selection (as shown in Chapter 6). This results in adjustments to environmental conditions that increase fitness (in terms of durability and efficiency and in the aggregate). By these means, ESS can be creative and adaptive. For instance, Sana Khan and Kirstie Fryirs of MacQuarie University in 2020 came up with methods to evaluate the "behavioral sensitivity" of rivers. Framing behavioral sensitivity as a dynamic property, they showed how rivers can evolve and shift between different sensitivity categories as they adjust to environmental change and disturbance.[8]

The advent of new forms or features can be major and global, such as the appearance of karst forms on carbonate rocks, of redox features as Earth's atmosphere became oxygenated, or the development of photosynthesis or backbones in biota. Novelties are often less dramatic and more localized—for example, emergence of new morphologies or flux patterns in geomorphology or biological speciation within existing taxa. Variation is a necessary precursor for novelty, in the form of deviations from the norm: irregularities, outliers, exceptions, deviations, and so on. If there are no variations, there is nothing to select for or against, and evolution cannot occur.

It is sometimes claimed that the abiotic environment is nonadaptive, and if one insists on adaptation directly analogous to biological adaptation, maybe so (this is why I prefer the term *adjustment* in reference to landscapes). Abiotic entities do, however, react to changes in inputs and boundary conditions, and certainly adjust or adapt in that sense, as we have seen.

In fact, a strong role for "ecological opportunity," often driven by abiotic environmental changes, in adaptive diversification of organisms has long been acknowledged and has been central to theories of speciation since Darwin. Ecological opportunity is defined as environmental conditions that allow the persistence of a lineage within a community and facilitate divergent natural selection within that lineage.

Niche availability occurs when a population with a phenotype previously absent from a community has the ability to persist within that community. Niche discordance occurs when an adaptive mismatch between a population's niche-related traits and the newly encountered ecological conditions generates diversifying selection. Ecological opportunity arises from both niche availability and discordance. Environmental change independent of the biota often produces ecological opportunities and adaptive diversification. For example, one study showed that sunfish in postglacial lakes experienced adaptive diversification due to the ecological opportunities created by geomorphic and hydrological changes. This followed work showing that postglacial dispersal, landscape features, and geomorphic barriers to migration were linked to genetic variability in bull trout in British Columbia. Plants are affected, too.

Dated plant phylogenies of twelve clades (groups of organisms descended from common ancestors) were found to reflect the impacts of climate and landscape evolution since the Neogene in the Cape Floristic Region of southern Africa. Substrate diversity increased due to geomorphic change, which spurred adaptive radiation.[9]

Evolutionary change can sometimes occur rapidly, on similar timescales to ecological processes, as shown by a review of studies of adaptive diversification and ecological opportunity. One example (fish again) is a study of bass in the Guadalupe River, Texas. Human-induced changes in river flows and land use/cover produced hydrological alterations that in turn led to systematic morphological variations in the fish; individuals from tributaries with different hydrological changes were morphologically distinct. Another study found that urbanization can drive evolutionary responses in fish in a few decades—though, naturally, the responses varied among different species.[10]

Landscape evolutionary creativity arising from biotic-abiotic feedbacks as life, landforms, and soils coevolve is supported by abundant evidence. Some key threads of inquiry include Dov Corenblit and colleagues' work on contemporary river systems, Neil Davies and Martin Gibling's studies of river systems in the geological record, William Verboom and John Pate's studies of plant-soil interactions in forests and woodlands, and Greg Retallack's studies of coevolution of flora, fauna, and soils in grasslands.

But what about evolutionary creativity independent of biota? This is explored below along three different tracks. Algorithmic evolution models exist that can explain evolutionary creativity in biota without reference to explicitly biological phenomena—and are applicable to abiotic components of landscapes. Second, the geological record reveals evidence of emergent novelty of landforms independent from biological evolution. Finally, mechanisms exist by which inheritable variations are produced independently of biota.[11]

Algorithmic Evolution

At first, it may sound a bit silly to propose a K-bit computer program as a metaphor for an organism. But the advantage of such a
simplified and generic representation is that whatever you discover
depends only on the program and is not affected by any of the
vast variability of real biota. Argentine-American mathematician
Gregory Chaitin, who pioneered this approach (his 2012 book
Proving Darwin is a readable introduction), proposed an organism
as a program that takes the original organism A and produces a
mutated organism A' (in this case, a modified version of the K-bit
program). Then some indicator of the adaptedness or fitness of the
of the organism/algorithm (could we call it an organithm or an algorism?) is computed—Chaitin used the largest integer the program
could calculate. That is, the larger the number, the better or fitter the
algorism. A mutation that produces a larger integer is kept (selected
for); a mutant that does not improve on the original is discarded or
selected against. This goes on until some maximum is reached. For
you (fellow) nerds, in Chaitin's models the limit is BB(N), defined
as the largest integer that can be produced by a $\leq N$-bit program
that produces a single integer and then halts. BB(N) stands for the
busy beaver number, a name you gotta love even though zoolo-
gists tell us that beavers are not really especially busy compared
to other mammals. These algorithmic models illustrate natural selection: random changes that do not improve fitness are discarded;
those that improve fitness are preserved.

Algorithmic evolution models show that evolution by natural selection is, or at least can be, an emergent process. A trend toward
increased fitness occurs from the select-or-reject criterion applied
to random variations, independently of any biological traits, environmental contexts, or guiding force. Such models are a reasonable first-order analog for biological evolution, where continuous
improvement is possible, up to some maximum limit (though in biology traits that do not improve fitness are sometimes retained). But

can analogous models or arguments work for, say, ecosystems or soils?

If changes in landscapes, whether due to internal dynamics, external factors, or both, affect fitness or adjustedness, the algorithmic evolution arguments apply to landscapes as well. And is there an appropriate measure of fitness in landscapes? As discussed earlier, most of the criteria proposed can be subsumed under efficiency or resistance selection—more efficient and resistant forms and structures are more likely to persist and grow, or to occur repeatedly, than less-efficient ones. And yes, there exist many metrics for measuring efficiency and resistance in ESS. If more efficient/resistant features and structures are subject not just to preferential preservation but also enhancement (via positive feedbacks) and/or propagation, then algorithmic evolution models apply to abiotic processes in landscapes, too—signifying the possibility of creativity.

But, but, but—while algorithmic evolution models climb toward the busy beaver number, and organismic modifications improve fitness in fits and starts, a possible objection to ever-increasing fitness or efficiency in, say, topography or hydrological flux patterns over long time periods is that these were achieved early on. That is, the very earliest flow networks, for instance, had the potential to develop optimal forms, even if not all did so. Some alluvial channels, for instance, evolving in the ways described by Nanson and Huang,[12] could perhaps have been optimally fit soon after their genesis.

But the geological and paleoenvironmental record also provides evidence of new, emergent forms and increasing variety as landscapes adjust to environmental changes. For example, except in highly cohesive material, most rivers were braided—until vascular plants and woody vegetation came on the scene. After such plants appear in the fossil record, so do meandering and anastomosing forms for river channels. These were new ways of maximizing efficiency, developed in materials where they were previously absent.

Creativity in biological evolution was accompanied by innovation in fluvial systems.[13]

As discussed in Chapter 6, selection for maximum efficiency has been shown to occur in many abiotic phenomena as well as in organisms and ecological systems. Selection as the persistence of more resistant forms is axiomatic in geomorphology and pedology. For evolutionary creativity to emerge in algorithmic evolution models, there need only be a selection process operating on new forms, assuming the new forms influence adaptations or adjustments to the environment. Therefore, implications of these models apply to abiotic and combined biotic/abiotic systems, subject to the appearance of new forms.

Note that maximum efficiency states are often non-singular; there is more than one way to achieve a given state. Maximum flow efficiency configurations in stream channels, for instance, can be reached with an infinite combination of quantitative values of width, depth, velocity, slope, and roughness, and even with multiple possible qualitative combinations of increases, decreases, or no change in the key hydraulic variables (something I worked on back in the early 1990s). There are obvious parallels to biological adaptations—think, for example, of the different adaptations organisms have to extreme cold or heat or to water excesses or limitations.

Novelties Appear!

Novelties have appeared, and continue to do so, in biota. For this reason alone, novelty also arises in communities and ecosystems. So, we need not belabor this point but rather focus on the appearance of novelties in other aspects of landscapes.

Individual ESS can, through various changes, generate ever more efficient variations until maximum efficiency is reached. Nanson and Huang describe how iterative adjustments accomplish this in channels. Biological evolution has produced similar innovations in different places and times (convergent evolution), and abiotic systems can also do so repeatedly as disturbances disrupt the efficient forms and structures.

No two dunes, sinkholes, topsoils, and so on are ever exactly the same. Singular, idiosyncratic aspects, by themselves, are not convincing evidence of emergence of new varieties. Evolutionary creativity in landscapes requires new elements to exhibit some qualitative or functional differences, or new modes of genesis, rather than just variations in detail. For example, overwash-resisting vs. overwash-absorbing barrier island morphologies or reflective vs. dissipative beaches are qualitatively and functionally different. Quantitative variations in, say, dune height, beach face slope, or longshore currents are not (though they are often related to the qualitative variations). For another example, variations in hydraulic conductivity in a soil pedon or layer are not, by themselves, functionally distinct. However, a transformation from a soil dominated by Darcian flow-through-porous media to one dominated by preferential flow paths and macropore flow qualifies as functionally distinct.

Entirely new geological entities have appeared in Earth history, associated with major events such as mass extinctions, global-scale tectonic events, oxygenation of the atmosphere, and the onset of glaciations. But can new forms be produced by the dynamics of landscape systems, independently of external factors? Nanson and Huang (those guys again) pointed out in 2018 that evolution is possible in abiotic systems with no genetic role (see, again, Chapter 6). If variations exist that differ in their functionality (efficiency, resistance, fitness), and if there exist selection mechanisms acting on these, then abiotic evolution is not only possible but also inevitable.

Abiotic landscape components cannot *reproduce*, but they can *produce*. Through a variety of geophysical and geochemical processes, abiotic systems can generate new forms. For example, long-term erosion may produce valleys from anticlinal ridges, and river downcutting can transform floodplains into terraces. Indeed, geomorphic change is essentially the production of new forms via transformations of earlier ones (Figure 9.2).

Figure 9.2. Landscape transformations revealed in the banks of the lower Guadalupe River, Texas. Arrows show approximate tops of at least three buried soil profiles, indicating transformation from a stable, very slow accreting surface (allowing soil formation) to a rapidly accreting floodplain.

Like biota, landforms and other abiotic features can also *propagate* or expand their geographic range or spatial coverage independently of size increases of individuals. This is most apparent with respect to the spread of, for example, certain types of vegetation or microbes. Landforms, soils, and hydrological entities expand spatially over large areas by the growth of individual features (e.g., fans, floodplains, deltas) but can also propagate via extension of certain modes of development (e.g., bedform migration, badland erosion, channel network expansion) (Figure 9.3).

Landscapes and their components can *die*—literally in the case of biotic elements and by being erased or obliterated in the case of abiotic. However, only organisms *must* die over timescales of a few centuries or less. Landforms and soils may persist beyond the lifespan of any organism. Sergey and Vladimir Krivovichev of Russia's St. Petersburg State University, along with Robert Hazen, showed in 2018 that minerals apparently do not go extinct. They analyzed almost 9000 data sets on the chemical composition or crystal structure of minerals and found evidence of an overall increasing trend of structural complexity. But, while more complex minerals have formed over geological time, simpler ones persist and are not replaced. Expanded analyses of even larger databases

Figure 9.3. Propagation of bedforms in the Sabine River, Louisiana/Texas. Note the smaller ripple forms superimposed on the larger sandbars.

have confirmed the trend of increasing complexity and preservation of older mineral lineages. Australians Stuart Mills and Andrew Christy in 2019 found that while minerals can temporarily disappear, they can reappear when physiochemical conditions are appropriate. The reversibility of "extinction" and the lack of direct dependence of successor or precursor minerals indicates, not surprisingly, that mineral evolution is not a direct analog of biological evolution.[14]

Landforms and soils associated with organisms that have gone extinct may themselves go extinct if no successor organisms with similar biogeomorphic or pedologic effects emerge. Abiotic features can become locally or regionally extinct due to burial, erosional removal, or geomorphic transformations. Global extinction is also possible—for example, weathering or other geochemical features linked to atmospheric or ocean chemical compositions that no longer exist. But can abiotic features go extinct (as opposed to local erasure or transformation) simply because they are less "fit," being supplanted by more efficient or durable forms? Though one could argue that is the case with resistance selection, when weaker forms or materials are removed, this is otherwise an open question.

Inheritable Variability

Four mechanisms may allow perturbations to persist and be inherited in landscapes. One is historical contingency, discussed in Chapter 4. The others are biological speciation effects via ecosystem engineering and niche construction, geological inheritance, and local disturbances.

Biological speciation can be associated with the appearance of new landforms or soils if the organisms have biogeomorphic ecosystem engineering effects or are bioconstructors. For example, the appearance of calcite-secreting cyanobacteria, coral organisms, and termites led to the appearance of stromatolites, coral reefs, and termite mounds. Other examples of biological evolution shaping the geologic environment include the advent of biomineralization genes and of bioturbation in the Cambrian, and the evolution of lignin in the Paleozoic, which stimulated the deposition of coal and various biogeomorphic impacts of woody plants.[15]

Variability of geology, coupled with perturbations, can also lead to inheritable variation. Consider geological effects on soils, groundwater fluxes, and landforms. Geological formations are given local and regional names because even within a given lithology, no two shale, sandstone, schist, or granite formations are identical. Considerable structural and lithological variation occurs within formations as well, which often profoundly influences resistance, permeability, landscape geochemistry, topography, and morphology—differences inherited from the geological variations. Consider also that inheritance can occur from abrupt geological

perturbations (e.g., earthquakes, tsunamis, and volcanic eruptions). These effects can be dormant and manifest later, as inherited geological variations appear—sometimes very gradually,
sometimes suddenly—as denudation proceeds, exposing new materials and structural features.

There exist different perspectives on geomorphological inheritance in the literature, but all are consistent with the notion of
passing on of traits as landscapes evolve. Twidale refers to *lineage* in addition to direct inheritance—that is, genesis of some
landforms can be traced to events that took place in the distant
geological past. He argues that many landforms are multistage and
have causative chronologies dating well back into time—even into
the Precambrian (Figure 9.4).[16]

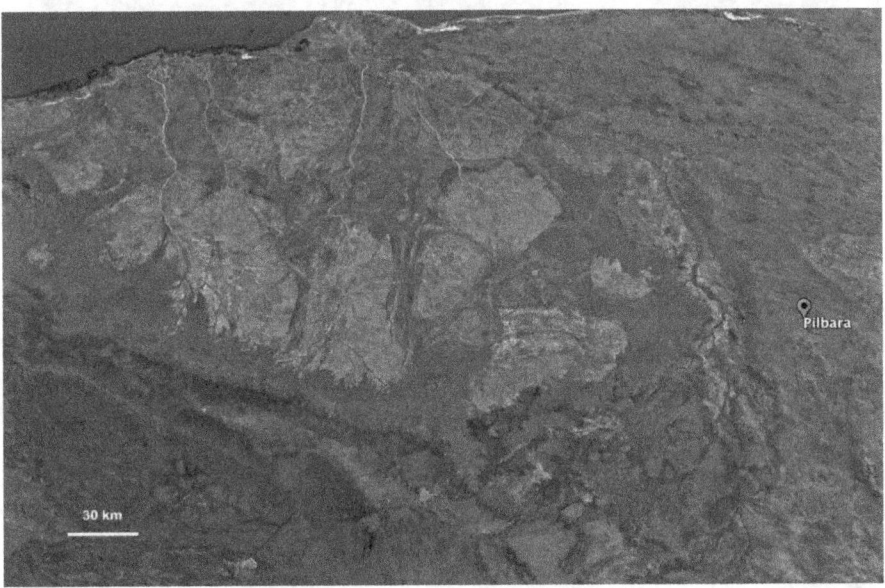

Figure 9.4. GoogleEarth[TM] image of the Pilbara Craton, Western Australia, the
oldest landscape on Earth.

Effects of local disturbances may also persist in the form of new
varieties of soils or landforms. These include storms, floods, fires,
and human agency. Some of these features that are new on a local
scale may also be novel in general (Figure 9.5).

Figure 9.5. Local disturbances—land clearing and grazing in a highly erosion prone setting—created these unique landforms: mega-gullies on the North Island of New Zealand. In the foreground at the bottom of the photograph is the channel of the Waipaoa River, filled with sediment from such features over the past century or so. This is an example of the "badass gullies" referred to earlier; see endnote 2.

Contingency as a source of variety refers to the fact that landforms and other ESS have irreducible elements of historical and geographical contingency. That is, no two places or environments have the same combination of environmental controls or influences or the same history of changes and disturbances (Figure 9.6) Even without appealing to landform individuality, these unique, idiosyncratic aspects of ESS consistently produce new varieties of landform, particularly if we endeavor to describe or designate landforms with greater specificity than just, say, gully or scree slope.

Figure 9.6. Left: though shell beaches occur elsewhere, the "perfect" combination of nearshore molluscan assemblage, and topography and wave and current energy regime, results in this unique feature along the South Carolina coast. Right: a unique combination of geology, landscape history, and erosional and weathering processes produced this distinctive mushroom rock in Egypt's White Desert.

What's in a Name? Taxonomy and Novelty

Identification of novelty may partly depend in part on how closely one looks. The main motivations for classifications, and main drivers of diversification, differ in biology as opposed to geomorphology, landscape ecology, geology, or pedology. The disparate levels of taxonomic detail in biology vs. other disciplines makes comparisons difficult. However, the disparity also suggests that recognition of new forms is strongly dependent on taxonomy. A lot of effort has gone into identifying the variety of life forms. There is currently extant only one species of the genus *Homo* and eleven of *Canis*, for instance. But there are more than 600 species of *Quercus* (oak trees) and >350,000 of beetles (and counting). ESS are categorized at higher levels of aggregation, so that soil types, landforms, and so on are more commensurate with, say, biological communities than with species. More than a century of soil classification and mapping, for instance, has so far identified <20,000 different soil types (series) in the United States.

Landforms are classified in even less detail than soils (though there exists no generally accepted scheme of geomorphic taxonomy or landform mapping, and the extent to which this represents greater natural variability of soils is unclear) (Figure 9.6). Works on aeolian geomorphology, for instance, list perhaps a couple of dozen types of sand dunes, while a vastly larger variety is evident even with my limited observations, considering sizes, shapes, activity, behavior, composition, soil and vegetation development, and subsidiary features such as blowouts. This limits efforts to inventory the appearance of new varieties.

EVOLUTION OF LANDSCAPE DIVERSITY

Here we consider the example of dolines (karst sinkholes, Figure 9.7). While karstification involves biological sources of CO_2 to help form carbonic acids, any CO_2-respiring organisms can perform this function. Therefore, while dolines are not abiotic, they are not linked to or dependent on any specific type of organisms and can evolve independently of biological evolution.[17]

Related to the taxonomic issues above, geomorphologists generally recognize just four to six basic types, depending on whether they are formed primarily by collapse into a subsurface dissolutional cavity or by dissolution from the top down and the presence or absence of a layer of regolith overlying the dissolving rock. But far greater variety exists.

Figure 9.8 is based on two principles: before dolines can exist, the conditions enabling a karst environment must be present, and there has to be a first time for everything. That is, the contemporary range of karst features did not appear simultaneously. From the top of Figure 9.8 down to "karst," the figure is chronological, according to the first principle above. Below that, it identifies

Figure 9.7. A selection of dolines. (A) Cenote II Ka, Yucatan, Mexico; (B, C) recently collapsed dolines in Florida and Kentucky; (D) Haiti; (E) doline and underlying cave passage exposed in roadcut, Kentucky.

some of the variation that exists, emphasizing dolines I am most familiar with in Kentucky—regolith covered, collapse-type, approximately circular in planform, drained by karst conduits, and often connected by underlying conduits or cave passages. Fat

arrows in the figure show many other pathways that could be followed, leading to many more types of dolines (and other karst
landforms). The bottom indicates *some* of the other ways dolines can differ. Many other possibilities exist, even for collapse
dolines.

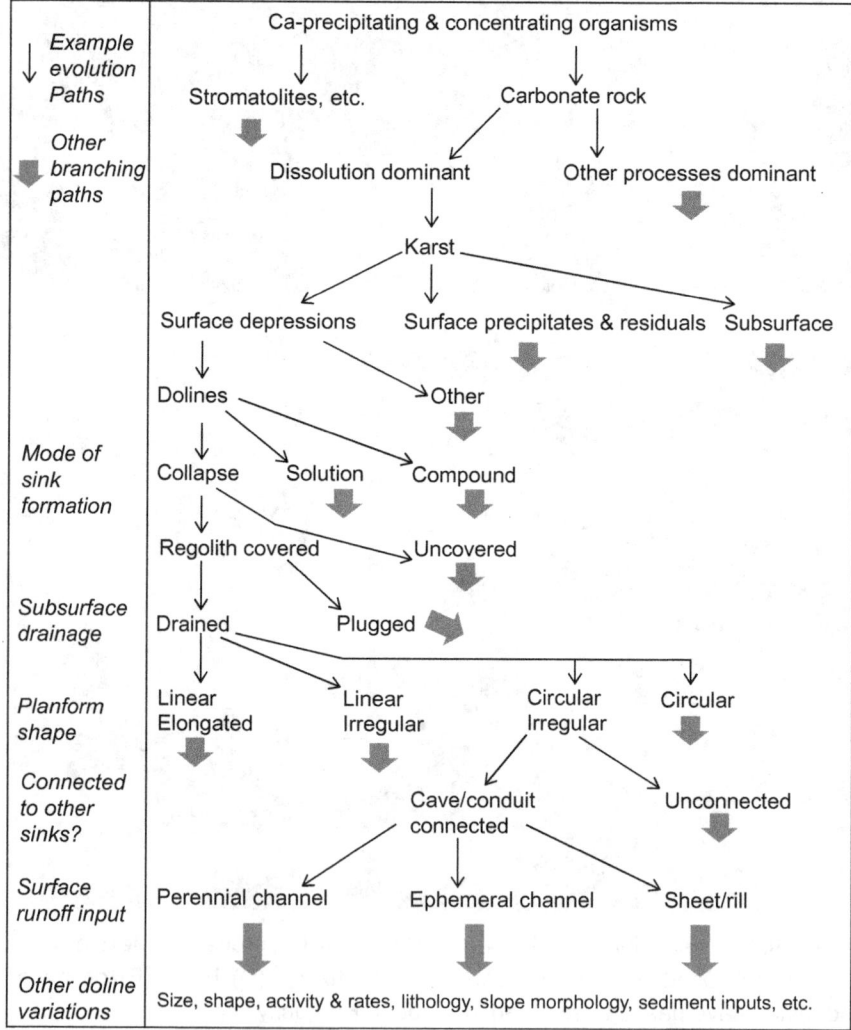

Figure 9.8. Evolution of dolines. In addition to the implied local variations
in geology, topography, hydrology, and so on, different locations would also
experience unique local histories of climate, weather, disturbance, biotic effects, etc. Source: J. D. Phillips, "Evolutionary Creativity in Landscapes," Earth
Surface Processes and Landforms 45 (2020): 109–120, Figure 6. Used with
permission.

For example, consider chronosequences on strath terraces (erosional terraces formed as the river incises) of the Kentucky River
(Figure 9.9). Dolines grow larger and often coalesce into com-
pound features and uvalas (a large depression formed by merging
of dolines) as the fluviokarst landscape develops. Plugging and infilling becomes more common in older dolines, but some sinkholes
show reversibility with respect to plugging. This occurs when major
runoff events flush out a drain swallet (which generally gradually
refills with sediment). However, once sedimentation in the doline
reaches the point where it becomes a (more or less) flat-bottomed
depression, the flushing-and-plugging ceases. Sometimes river
flow takes a shortcut across the meander bend via joint-controlled
groundwater conduits. Enlargement of the latter can trigger collapse of overlying material, expressed at the surface as a nearly
straight line of dolines. These represent a new formation mechanism not present on the upland. Alluvial (depositional) terraces, as
opposed to strath terraces, are not preserved, so they are present
only in the valley bottom. As the meanders grow, alluvial collapse
dolines formed on the valley bottom are not preserved. Therefore,
there are examples of new forms emerging and of local extinction.

Selection is most apparent as efficiency selection in flow paths,
involving occupation or formation of conduits and channels, reinforced by erosional feedbacks. Preferential preservation of the
least jointed and fractured rock masses is a manifestation of resistance selection. Efficiency selection shapes the morphologies of,
for example, fluvial channels, karst conduits, and dolines. Additionally, structural and lithological control may obscure other factors.
Greater fitness is manifested in the Kentucky doline case study as
more efficient landscape-scale hydrological flow patterns as well
as increasing prominence of resistant features.

This case study shows creation of new doline forms as well
as the vanishing of forms as the landscape evolves. But "extinction" of alluvial collapse dolines is only local, as new ones can
form, and similar features exist elsewhere. This raises the question of whether any dolines in the Kentucky River study area

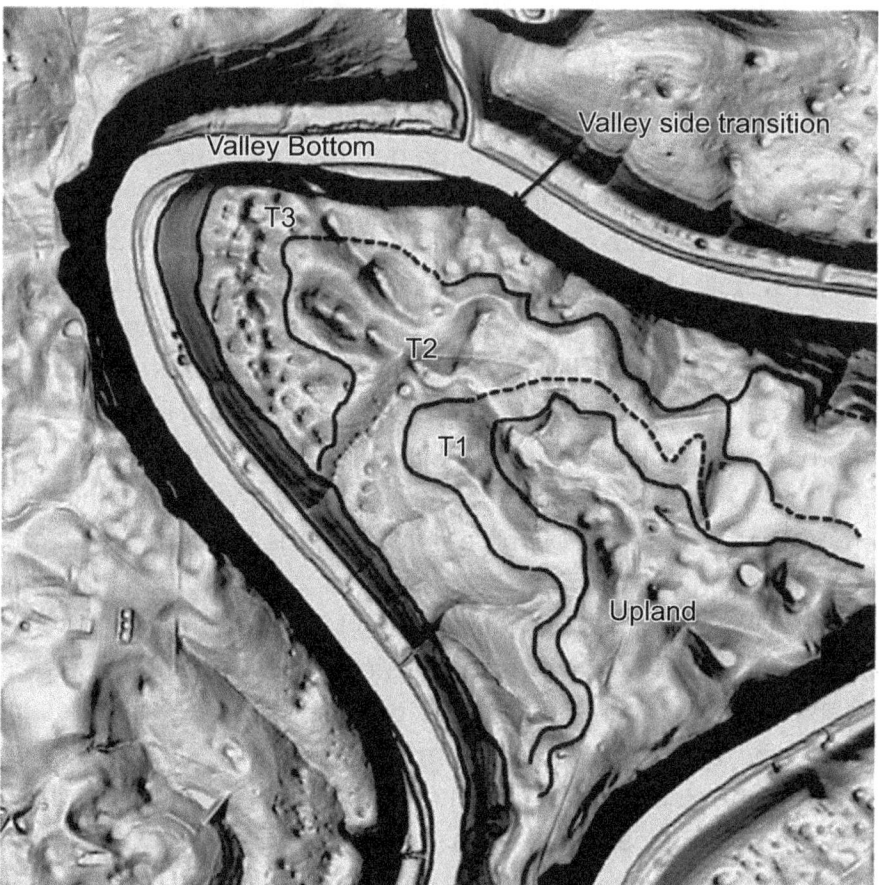

Figure 9.9. Chronosequence at Polly's Bend on the Kentucky River. The upland is the oldest, most highly developed karst area, with strath terraces T1, T2, T3, and the valley bottom successively younger. Base map is shaded relief based on 1.5 m resolution digital elevation model. Area shown is about 2.7 km (north-south) by 2.5 km (east-west). Coordinates at center are 37.8022° N, 84.6472° W. Source: J. D. Phillips, Historical Contingency in Fluviokarst Landscape Evolution," Geomorphology 303 (2018): 41–52, Figure 2. Used with permission.

J. D. Phillips, "Historical Contingency in Fluviokarst Landscape Evolution," *Geomorphology* 303 (2018): 41–52.

differ in any meaningful way from dolines elsewhere. Unfortunately, we can't be sure because the anatomy of dolines is not systematically recorded in sufficient detail. However, there is evidence that the setting driving doline evolution is specific to the Kentucky River meander bends. Development of incised meander bends featuring dense karst development on bend interiors, along with strath terraces representing a chronosequence, has not been reported in the research literature. Incised meanders formed in thick layers of relatively pure limestone such as those in the study area are uncommon to begin with. Where they are found, place factor variations (e.g., hydrologic regime of the fluvial system, composition and structure of the carbonate rocks, regional topography, and distance above base level) are present. So are differences in historical factors such as the rate, duration, and timing of river downcutting. Together these create important differences. In an earlier study, I examined topographic maps of meandering rivers in other limestone regions of the United States to see if similar morphological features exist. The answer is no, save for a short reach of the Green River, Kentucky—but even those sites showed some meaningful differences from the Kentucky River.[18]

SUMMARY

Landscapes reflect a specific combination of applicable laws and historically and geographically contingent factors that are extremely unlikely to be duplicated elsewhere. This is the core of the *perfect landscape* concept. As more and more variables are accounted for when comparing landscapes, their individuality (or perfection) can only increase. Nine axioms for landscape interpretation can be applied to understanding perfect landscapes, based on principles of spatial structuring, selection, coalescence, constrained individuality, mutual adjustments, canalization, constant change, reversibility, and scale dependence.

Evolution of perfect landscapes implies creativity: the appearance of new features that are selected for. Such creativity is evident in biological evolution but also occurs in abiotic phenomena. Contemporary empirical studies and the geological record show examples (such as the case study of dolines above) of the emergence of novel forms: geophysical geochemical, and abiotic-biotic hybrids as well as in organisms.

NOTES

1. J. D. Phillips, "Human Impacts on the Environment and the Primacy of Place," *Physical Geography* 22 (2001): 321–332; J. D. Phillips, "The Perfect Landscape," *Geomorphology* 8 (2007): 159–169.
2. J. D. Phillips, "Badass Geomorphology," *Earth Surface Processes & Landforms* 40 (2015) 22–33.
3. M. Marden et al., "Badass Gullies: Fluvio-Mass-Movement Gully Complexes in New Zealand's East Coast Region, and Potential for Remediation," *Geomorphology* 307 (2018): 12–23; I. C. Fuller et al., "Badass Gully Morphodynamics and Sediment Generation in Waipaoa Catchment, New Zealand," *Earth Surface Processes and Landforms* 45 (2020): 3917–3930.
4. "Inherent complexities and uncertainties prompt perceptions of the process of interpretation in geomorphology as a frustrating form of witchcraft or wizardry—a dark art."—G. J. Brierley, K. Fryirs, H. Reid, and R. Williams, "The Dark Art of Interpretation in Geomorphology," *Geomorphology* 390 (2021): 107870.
5. J. D. Phillips, "Place Formation and Axioms for Reading the Natural Landscape," *Progress in Physical Geography* 42 (2018): 697–720. Brierley et al. (2021, endnote 4) developed a compatible but quite different practical approach to landscape interpretation relating generalized understandings derived from remotely sensed data to field observations and measurements and local knowledge, supporting contextualized place-based applications. They called their cognitive approach "Describe-Explain-Predict," whereby explanation builds upon meaningful description, leading to reliable predictions based on multiple lines of evidence.
6. K. M. Johnson and W. B. Ouiment, "An Observational and Theoretical Framework for Interpreting the Landscape Palimpsest through Airborne LiDAR," *Applied Geography* 91 (2018): 32–44; L.-C. Hsu and J. A. Stallins, "Multiple Representations of Topographic Pattern and Geographic Context Determine Barrier Dune Resistance, Resilience, and the Overlap of Coastal Biogeomorphic Models," *Annals of the American Association of Geographers* 110 (2020): 640–660.
7. T. de Haas et al., "Holocene Evolution of Tidal Systems in the Netherlands: Effects of Rivers, Coastal Boundary Conditions, Eco-Engineering Species, Inherited Relief and Human Interference," *Earth-Science Reviews* 177 (2018): 139–163.
8. S. Khan and K. Fryirs, "An Approach for Assessing Geomorphic River Sensitivity across a Catchment Based on Analysis of Historical Capacity for Adjustment," *Geomorphology* 359 (2020): 107135.
9. W. M. C. Jarvis, K. R. Peiman, S. J. Cooke, and B. W. Robinson, "Low Connectivity between Sympatric Populations of Sunfish Ecotypes Suggests Ecological Opportunity Contributes to Diversification," *Evolutionary Ecology* 34 (2020): 391–410; A. B. Costello et al., "The Influence of History and Contemporary Stream Hydrology on the Evolution of Genetic Diversity within Species: An Examination of Microsatellite DNA Variation in Bull Trout, Salvelinus Confluentus (Pisces: Salmonidae)," *Evolution* 57 (2003):

328–344; V. Hoffman, G. A. Verboom, and F. P. D. Cotterill, "Dated Plant Phylogenies Resolve Neogene Climate and Landscape Evolution in the Cape Floristic Region," *PLoS One* 10 (2010): e0137847. https://doi.org/10.1371/journal.pone.0137847; a recent review of adaptive radiation studies in plants is provided by J. J.Schenk, "The Next Generation of Adaptive Radiation Studies in Plants," *International Journal of Plant Sciences* 182 (2021): 245–262.

10. J. T. Stroud and J. B. Losos, "Ecological Opportunity and Adaptive Radiation," *Annual Review of Ecology, Evolution, and Systematics* 47 (2016): 507–532; J. E. Pease, T. B. Grabowski, A. A. Pease, and P. T. Bean, "Changing Environmental Gradients over Forty Years Alter Ecomorphological Variation in Guadalupe Bass *Micropterus treculii* throughout a River Basin," *Ecology and Evolution* 8 (2018): 8508–8522; E. M. A. Kern and R. B. Langerhans, "Urbanization Drives Contemporary Evolution in Stream Fish," *Global Change Biology* 24 (2018): 3791–3803.

11. The evolutionary creativity discussion here is derived from J. D. Phillips, "Evolutionary Creativity in Landscapes," *Earth Surface Processes and Landforms* 45 (2020): 109–120.

12. Referenced in Chapter 6.

13. This has been demonstrated in laboratory and numerical models and is consistent with field evidence. Most convincingly, there exists geological and paleontological evidence of coevolution of meandering rivers and woody vegetation. See, for example, N. S. Davies and M. R. Gibling, "Cambrian to Devonian Evolution of Alluvial Systems: The Sedimentological Impact of the Earliest Land Plants," *Earth-Science Reviews* 98 (2010) 171–200; N. S. Davies and M. R. Gibling, "The Sedimentary Record of Carboniferous Rivers: Continuing Influence of Land Plant Evolution on Alluvial Processes and Palaeozoic Ecosystems," *Earth-Science Reviews* 120 (2013): 40–79; M. R. Gibling et al., "Palaeozoic Co-Evolution of Rivers and Vegetation: A Synthesis of Current Knowledge," *Proceedings of the Geologists' Association* 125 (2014): 524–533. Note that environments dominated by fine-grained, cohesive sediments are an exception—that is, they can have sufficient bank strength to allow meandering patterns without vegetation reinforcement.

14. S. V. Krivovichev, V. G. Krivovichev, and R. M. Hazen, "Structural and Chemical Complexity of Minerals: Correlations and Time Evolution," *European Journal of Mineralogy* 30 (2018): 231–236; S. J. Mills and A. G. Christy, "Mineral Extinction," *Mineralogical Magazine* 83 (2019): 621–625.

15. D. J. Bottjer, "Geobiology and Palaeogenomics," *Earth-Science Reviews* 164 (2017): 182–192.

16. Twidale's arguments are laid out in C. R. Twidale, "The Enigma of Survival: Problems Posed by Very Old Paleosurfaces," *Physical Geography* 24 (2003): 26–60; and C. R. Twidale, "Lineage as a Factor in Landscape Analysis," *Physical Geography* 26 (2005): 23–51. Other influential perspectives on geological and geomorphological inheritance include P. W. Williams, "Geomorphic Inheritance and the Development of Tower Karst," *Earth*

Surface Processes and Landforms 12 (1987): 453–465; and F. Ahnert, "Equilibrium, Scale and Inheritance in Geomorphology," *Geomorphology* 11 (1994): 125–140.

17.   This case study is based on Phillips (2020); endnote 11.

18.   The earlier study referred to was called "The Badass Bends of the Kentucky River," included in the reference in endnote 2.

10

Climate Change and Mysterious Ways

DO SOMETHING!

As climate change and its many impacts unfold, many worse than we had forecast or feared, it is often said that Earth is entering a "new normal." This is not wrong. However, with respect to our ability to understand, adapt to, and predict environmental change from here on out, it is probably more accurate to say there is *no* normal. The climate and environment that we will contend with will be unlike any our species—much less our infrastructures, institutions, and cultures—has ever encountered. I agree with those who say, sometimes circumspectly and sometimes directly, that it is time to panic—not in the sense of panic as uncontrollable fear or anxiety that can cause wildly unthinking behavior, but in the sense of another definition: a frenzied hurry to do something. Scientists hate to be called alarmist, but when the house is on fire, you sound the alarm. From the previous nine chapters we can distill some guidelines for action, explained below and summarized in Box 10.1 at the end of the chapter. *Do something* involves taking whatever measures we can, as vigorously and as soon as we can, to cut anthropic carbon emissions and limit the damage. *Do something* involves recognizing that changes are underway, like it or not. To ask whether one believes in climate change is like asking whether they believe in cells or earthquakes—climate change,

Mysterious Ways. Jonathan D. Phillips, Oxford University Press. © Oxford University Press (2025).
DOI: 10.1093/9780197755129.003.0010

cells, and earthquakes exist, whether you or I believe in them or not. *Do something* involves figuring out how to adapt to change and mitigate its adverse impacts.

As bio- and geoscientists, we must come to grips with the fact that climate change is influencing, or soon will, *everything we study*—landforms, soils, ecosystems, water resources, agriculture, forestry, etc., etc. That doesn't necessarily mean that every geomorphologist, hydrologist, or ecologist should make climate impacts the focus of their work. It does mean that we should think about how what we know, think we know, and learn can inform our understanding of and adaptations to climate change and its impacts. Because *everything is connected to everything else* and *all other things are never equal*, we will have to confront more complexity. These principles have always applied, but global climate shifts change *everything*, at more or less the same time. Direct experience (i.e., drawing direct parallels between past and future) will be less helpful as we move into uncharted territory in terms of providing a template for predicting change, while the *lessons* we learn from the past about complex interactions will become even more important.

Much can be, has been, and will be learned from studying the closest precedents we can find in Earth history to the current and near-future situation—periods of higher or rising temperatures, greater carbon dioxide emissions, rising sea levels, etc. But with respect to effects on human affairs, we are quickly encountering situations and scenarios for which there are no precedents. And even though there were episodes of, for example, rising sea level or retreating glaciers in the past, those rising seas did not threaten developed coastlines, and those glaciers were not water supplies for numerous human settlements and economies. In addition to examining precedents, we should be looking for *analogs*. In many cases, these will be hotspots of recent and contemporary change that may be bellwethers for more general or global phenomena. For instance, lessons from places such as south Louisiana, where subsidence and other factors create a rate of relative sea level rise greater than the rate of eustatic sea level change, may be crucial in an era of accelerating global sea level rise. Rather than focusing on typical or average glaciers, studies of those retreating most rapidly may provide the most useful information for assessing future impacts of glacial decline.[1]

This may sometimes go against our training and traditions—after all, most of us were taught and conditioned to seek a holy grail of "representative" sites. We need to shift (or at least broaden) that focus from seeking out sites representative of what was going on until recently to locations representing the overheated, turbo-charged world of changes we find ourselves in.

We often speak of past and future in terms of historical *trajectories*—I have done so in this book. The formal definition of trajectory has to do with the path followed by an object moving under the action of given forces. The paths and velocities may be inconstant, and trajectories can be modified, but the definition implies that they are deterministic and covered by (presumably known) rules. The trajectories we've described and discussed here are historical developmental pathways. This is understood well enough, I think, but *trails* may be a better metaphor for ongoing and future change. While the rules or laws governing development of Earth systems may be constant, the boundary conditions and inputs are shifting rapidly. And the historically and geographically independent rules and laws are not the only thing influencing evolutionary pathways. Trails are influenced and partly constrained by invariant principles, but also by idiosyncratic, historically contingent controls, and are potentially affected by a variety of factors that have little or nothing to do with the physics of movement. The trail metaphor is more apt for the interpretive and predictive problems we face, particularly if (or as) critical thresholds are crossed (e.g., a sea ice-free arctic summer, ocean heat absorption, livability of overheated cities).

COMPLEXITY, COMPLEXITY, COMPLEXITY

Complexity can't be avoided and might as well be embraced. Some of the principles of complexity, nonlinear dynamics, and contingency have been interpreted as implying that (some) Earth surface systems (ESS) are intrinsically unpredictable. Not so much. What these principles indicate is not hopeless unpredictability but a *new context for prediction*.

I have often stressed uncertainties and multiple possibilities, but let's be clear: *uncertainty is no excuse for inaction*. When the house is on fire, you may be unable to estimate the rate of heat release, the flammability and consumption rate of various materials in the house (which will also vary with their age and ambient environment), the probabilities of fire propagation in various directions, or the fluid dynamics of air flow in the vicinity, which will of course change as the fire proceeds. You do know, however, how to put the fire out and that you damn well better call the fire department or grab a hose or bucket or extinguisher. You also know that it would be idiotic and disastrous to calculate or debate the cheapest possible way to contain the fire and the minimum amount of water required to quench it. The house is on fire, and it is past time to grab a bucket.[2]

Understanding, assessing, and predicting impacts of climate change is not at all like programming a computer, where a specific set of instructions produces a specific output (though even in this case deterministic chaos is possible). It is much more like sailing a boat, where a good sailor will make frequent adjustments to the wind and waters, based partly on ironclad universally applicable scientific principles, partly on general sailing principles and practices, and partly on instinct and intuition based on training and experience.[3]

The broader question, of course, is what we want to program the computer to do (or where we want to sail the boat). Unfortunately, we cannot now—and may never be able to—plot a precise course in advance. We can, however, identify some signposts and landmarks and rules of thumb for pathfinding—some stars to steer by.

The kind of research I advocate and practice is not intended to make specific predictions or forecasts but to identify some alternative paths. Not to devise global laws, but to present general lessons.

LESSONS FROM THE HOTSPOTS AND COLD SPOTS

The warnings about human-accelerated climate change we've been hearing (and those of us in the business have been sending) for decades are, unfortunately, coming true. Almost daily, our news feeds remind us of this or provide new evidence that Earth's climate, and the environmental systems affected by it, are approaching unknown territory. We are seeing ocean temperatures rise, ice loss from the great Antarctic and Greenland ice sheets, storm and flood regimes, heatwaves, and fires that are unprecedented in human history and in some cases unprecedented in Earth history, period.

Engineering design, insurance, land-use planning, and economic forecasting, among other things, are grounded on statistical analyses and risk assessments based on data from the past—"from the past" would ordinarily be unnecessary, as there are no data from the future. But I wish to emphasize that the data record is becoming increasingly irrelevant to the present and future. What used to be rare and extreme heatwaves, tropical cyclones, fire seasons, and so on are becoming commonplace. The hundred-year storm or flood concepts, for instance, essentially apply to the twentieth century. Beyond analysis of recorded data, we often rely on deeper historical information such as paleoclimate or paleoecological evidence to guide us in understanding environmental change. All this gives useful knowledge and clues about ongoing and future change. But we must temper our interpretations with the knowledge that the Earth system and its components ain't what they used to be—not just over geological timescales but also over human history and even the lifetimes of old folks like me.

If Earth history is not necessarily representative of what is happening now and likely to happen in the future, what are we to do? A potential answer is to identify and focus on *hotspots* and *hot moments* (not necessarily, though sometimes, in a literal thermal sense) where and when specific landscapes are or were responding to changes such as those that are now occurring. It's a simple concept and not new, but it is worth revisiting.

In studying impacts of heatwaves on urban environments, for example, look for the ones that have exhibited the most pronounced urban heat island effects or that have historically endured extreme heatwaves. Focus on ecosystems that were already fire-prone or frequently burned to gain insight into how more frequent fires may affect other ecosystems or what adaptations nature has come up with that we might mimic or encourage. Continue to look at events such as Meltwater Pulse 1A (about 14.2 ka) when rapidly melting ice raised sea levels at an average rate more than ten times greater than twentieth-century paces—not as an anomalous event but as a lesson for a possible near-future scenario.

The Waccamaw example from Chapter 8 suggests the importance of examining, if you will, *cold spots* (not at all in the literal thermal sense) where impacts of changes and disturbances likely in the future are readily absorbed. If you recall, from 2015 to 2018 the lower Waccamaw River experienced the three largest river flows and highest water levels on record. Two of the three storms included both unprecedented river discharges from upstream and storm surges from downstream. Yet, in the lower Waccamaw River fluvial-estuarine transition zone (FETZ), geomorphic impacts were minimal, in sharp contrast to other landforms and ecosystems in the region. The reason is that the FETZ is a complex of various types of channels, wetlands, and water bodies that is perfectly adapted to handle large volumes of water from upstream/downstream. This is because the FETZ has developed under the influence of rising sea levels during the Holocene, constantly subjected to stream discharge coming downstream and sea level gradually encroaching upstream. What aspects of such cold spots and cold moments can we facilitate or replicate?

The traits of the Waccamaw FETZ that enabled it to absorb these impacts include extensive wetlands, very high channel-wetland connectivity, various "spillways" for exporting and storage areas for storing excess water, and water flows and exchanges between these components that can move in multiple directions depending on circumstances. For the case of river, wetland, and coastal environments and their response to more and more powerful storms and floods and to sea level rise, this points to the importance of preserving, protecting, and perhaps restoring or rehabilitating wetlands and hydrogeomorphic features that facilitate the key traits and dynamics. It also suggests the importance of multiple degrees of freedom or ways to respond to changes.

By examining the hotspots of climate-driven change and the cold spots of resistance or resilience to climate change, we can, hopefully, gain insight as to what to expect and what to do as the climate change you-know-what continues to hit the fan.

RADicalism

The ecosystem management and applied ecology community has recently seen the rise of the RAD framework, for resist-adapt-direct. The framework was developed by and for ecosystem managers, mainly in US federal agencies such as the US Fish and Wildlife Service. The RAD framework was instigated by recent, ongoing, and projected ecological transformations driven by rapid climate change, and, as its proponents say, encompasses the entire decision space for managers. *Resist* refers to attempts to stop, slow, or reverse the changes. This is the traditional response and may still be appropriate or necessary when environment change threatens, for example, irreplaceable cultural resources or critical infrastructure. *Adapt* approaches involve accepting the changes, with careful monitoring to facilitate adaptations. The *direct* part encompasses efforts to steer changes toward desired and away from detrimental pathways and outcomes.[4]

The RAD concept recognizes the inevitability of environmental change, the dramatic and rapid tempo at which it is happening, and the fact that many changes will create unprecedented states of ecosystems and landscapes. RAD rejects resilience—not in general or completely, but as the overriding main goal of management, as resilience implies maintaining or restoring ecosystems in or to their current or some desired former target state. Resilience and resistance are still considered to be potential management goals or options, but RAD recognizes that they will not always be possible, feasible, or even desirable. Management has generally been (and mostly still is) based on history—for example, flood or storm probabilities based on statistics of the historical record, historical ranges of environmental variability, and restoration of historical system states. In the context of contemporary climate change, history is increasingly irrelevant as Earth moves into new normal or no-normal conditions of a constantly moving baseline. As such, resist-adapt-direct is a useful and welcome refocus of management. As geographer John Williams of the University of Wisconsin put it, "Even if anthropogenic climate change wasn't a thing, the paradigm of stationary historical baselines was long overdue for the scrapheap. Viewing the past as stable was always only a convenient assumption."[5]

Once the research or management problem is framed, RAD asks five general questions: First, is the transformation a threat and to what or whom? Values, interests, and objectives must be identified along with critical uncertainties. Question two is how durable and effective resistance strategies might be. In question three, possible futures are identified and narrowed down to those that are plausible and those potential outcomes linked to the values and priorities identified earlier. The fourth question is about the potential consequences of resisting, adapting, or directing change. Question five involves making choices in the RAD decision space. The framework also assumes monitoring, observation, and learning to feed back to the cycle of questions and decisions.[6]

A team from the US Fish and Wildlife Service and other federal agencies gives examples from Blackwater National Wildlife Refuge in Maryland, which is undergoing rapid modification driven by relative sea level rise and climate change. In the marsh and wetland-dominated refuge, a resistance strategy is in place for a portion of the refuge that provides critical habitat for endangered species. This involves measures such as adding thin layers of sediment to help wetlands keep pace with coastal submergence. An accept/adapt approach is applied to areas invaded by non-native *Phragmites australis* (common reed). Though refuge managers consider *Phragmites* suboptimal in terms of the refuge's primary goals of providing wildlife habitat, the reed is notoriously difficult and expensive to remove or control. Managers have no funding to address this, so accepting it was the only option. A "direct" example occurs elsewhere at Blackwater, where tree removal encourages tidal marsh movement into upland forest, which is occurring more slowly anyway, to replenish the marsh habitat considered most critical for refuge goals.[7]

As difficult as it might be to devise tactics for resisting, accepting, or directing, decisions about which RAD strategy to pursue pose the biggest potential problems. Science, engineering, planning, and management already provide many of the tools for *how* to R, A, or D and have a pretty fair track record of developing or improving methods where this is not the case. But *what* to R, A, or D or even *whether* to do anything brings politics into play, along with the associated political power dynamics, economic, and social factors. Our scorecard as a species on issues such as carbon emissions, pollution control, and environmental cleanups does not inspire confidence.

Will decisions on what to prioritize for R, A, or D be guided by humanist concerns, scientific principles, and data? Or will they be made based on the wishes of rich and powerful elites or on the ability to pay? If you choose to build a home in an area of high fire risk or on a barrier island, will we (continue to) subsidize that choice with infrastructure, insurance, and emergency response paid for by society as a whole? Not all of us have the professional expertise to address this at a research or tactical policy level, but we all have opportunities to influence strategies as citizens, voters, consumers, and personal economic decision-makers.

Now we turn attention to some of the lessons the mysterious ways of nature teach us that can help us adapt to environmental change or at least comprehend what the heck is happening to us.

HOW I STOPPED WORRYING ABOUT CONTINGENCY AND LEARNED TO LOVE IT

The state of an ESS is a function of generally applicable laws that ultimately determine the range of possibilities, geographically specific place factors (environmental constraints and opportunities), and history. While laws are general, if not universal, and apply to every ESS of a given type (e.g., stream channel, cave, mangrove swamp, soil profile), the place factors define the template in which those laws operate.

This is geographical contingency, which means that the answer to most questions about how landscapes respond to climate change starts with "it depends . . . ." That can be frustrating for those of us trained, as I was, to constantly strive toward unequivocal answers with minimal uncertainty. Traditionally, most of us tried to beat contingency into submission by collecting ever more, and ever more detailed, data and by introducing more variables or processes into our models. Nothing wrong with that, but as discussed in Chapter 9, considering more factors or variables can only increase the uniqueness or idiosyncrasy of our representations. As Columbia University's Ruth DeFries put it, nature teaches us that "one size fits none."

And then there is history.

The patch of land where I sit now, for instance, in its current state is contingent on any number of past climate changes, sea level oscillations, geological events and episodes, biogeographic dispersals, and biological speciations or extinctions (thus, at longer timescales, even the place factors have elements of historical contingency). On shorter timescales, it depends on tropical storm climatology; the magnitude, timing, severity, and specific tracking of individual tropical cyclones; initial conditions at the outset of these cyclones; and other events and their specifics such as floods, tornadoes, severe rainstorms, and extratropical cyclones. It depends on whether lightning strikes or a cigarette is tossed on the ground (and whether either ignites a fire), the fuel conditions on the ground, meteorological conditions when a fire ignites, and in the past couple of centuries, whether humans decided to put it out or let it burn. The state of this ESS depends on land ownership and management and an uncountable number of decisions and happenstances. Due to teleconnections, the historical contingencies are not even confined to things that happen (or not) here. Climate and sea level effects of changes in Antarctic ice, sea surface temperatures in the tropical Pacific, and timber prices in India or Brazil, for example, may all leave their mark.

And so it is wherever you are right now, wherever you have been, and wherever you may go. The details are always different, but the phenomenon of historical contingency is always present.

If you look back at a known or hypothesized time zero starting point in terms of what could have happened, or what would happen if you could rewind the clock and start over, there are multiple possible pathways and outcomes at any given point in time. If you think about it in sufficient detail, there may exist infinite possibilities. Note that infinite does not mean that anything could happen—there are an infinite number of even integers, for example, but it is not possible for that set to include non-integers or odd integers.

If you represented this as a mathematical directed graph with system states as the nodes and possible changes through time as the links, even if there were just two possibilities at each juncture or point of potential change, there would be 2^q possible states after q episodes of change (storms, lightning strikes, land-use decisions, etc.).

However, if you look at the state of an ESS at any given time, the chain of events and state changes that led to that would be a linear sequential directed graph—that is, A led to B, B to C, and so on. From this perspective, the observed ESS is a singular outcome of a particular chain of events. Even if you acknowledge that the hypothetical resetting of the clock or rewinding of the tape would not produce the same result, history is irreversible.

So, from the standpoint of an observed or experienced ESS looking back, it is possible that multiple pathways *could* have led to what you've got, but only one pathway *did*. It is certain that other trajectories could have led to different outcomes from the same starting point, but only one trajectory and one outcome occurred (Figure 10.1).

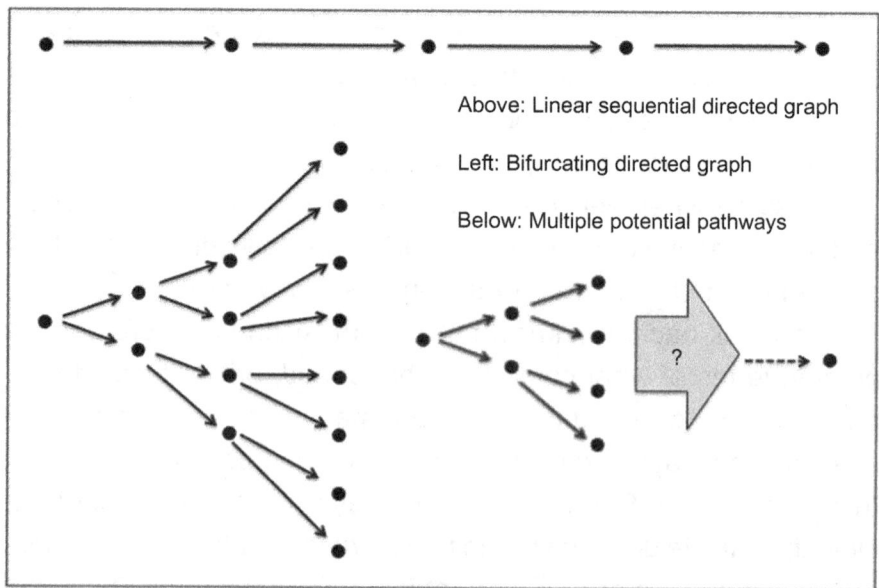

Figure 10.1. ESS evolutionary sequences. Top: what did happen. Bottom left: what could or could have happened. Bottom right: typical interpretive situation, with uncertainty about how an observed state did happen, and multiple possibilities.

Looking ahead, multiple future trajectories and outcomes are possible. Those can be constrained based on law and place factors, but the possibilities become larger and essentially limitless further into the future. Yet, of course, only one trajectory will occur, with each change opening new possibilities but foreclosing other possible pathways.

Four Sure

$$2 + 2 = 4$$

That is non-contingent. Adding $2 + 2$ gives the same result no matter who does it, how they do it, where they do it, or when. The same goes for expressions such as $2 + X = 4$, or $27/X = 4$, etc. This four-play is a metaphor for the deterministic, Laplacian, non-contingent ideal of science, where the right tools and sufficient information always give the same, correct result under any circumstances.

But a better metaphor for the actual practice of geosciences and other historical, field-based sciences is that you find or observe (evidence of) a four—or of a number between 3.5 and 4.5. Mathematically, of course, there are an infinite number of numerical operations that could produce a four. Even if you know that the four arose from, say, subtraction or exponentiation, the possibilities are still infinite.

Sometimes we can discover clues or information that allow us to construct something like an algebraic expression and make the problem more deterministic. More often, however, we are faced with constraining, somehow, the infinite possibilities.

For a simple but realistic example, think of stream discharge (Q), which (using the Chézy equation) is given by

$$Q = w\, d\, C\, R^{0.5}\, S^{0.5}$$

where w and d are width and mean depth, C is the Chezy coefficient (which combines some physical constants with a roughness or friction factor), R is hydraulic radius (cross-sectional area divided by wetted perimeter), and S is the energy grade slope.

You have an observation of Q, and you want to know the properties of the flow that produced it (these kinds of problems come up in paleohydrologic reconstruction as well as in studies of flow hydraulics). Even if you assume, as is often the case and with some validity, that $R \approx d$, you have five variables to consider. In terms of the arithmetic, there are an infinite combination of values of w, d, C, and S that could produce a given Q. But you can constrain the infinite possibilities; you must if you hope to make a respectable stab at the problem.

Presumably you have some idea of the dimensions of the channel. Based on the banktop width, you can (assuming the flow was not an overbank flood) eliminate values of w higher than this and perhaps also identify implausibly low widths. Similarly, you may be able to constrain plausible depths based on local base levels, the topographic setting, and geotechnical constraints on bank heights in a given material. While S is not the same as channel bed slope, it is pretty close in some cases, and you can often identify a range of possible energy grade slope values.

Of course even if you know, for instance, that width had to be somewhere between 5 and 15 m, and depth <5 m, there are still an infinite number of values between 5 and 15 or 0 and 5. Commonsense considerations can make this finite, for example by limiting the values to 1, 2, or 3 decimal places (who measures in the field to the nearest millimeter, anyway?). But you still often have multiple, sometimes many, possibilities.

Mathematicians and philosophers have long pondered, and continue to debate, whether infinity is a concept with real physical meaning or simply a useful mathematical abstraction. Whether there are truly unlimited numbers of possible stream channel geometries, vegetation compositions, or whatever, there are clearly in our business some cases where the number of possibilities is *essentially* infinite, in the sense of having no known limit or far more possibilities than we can deal with—or that could occur. Imagine some type of bifurcating system of sixty iterations. As the number of seconds elapsed since the Big Bang is about 2^{59}, they'd have to occur at a frequency of >1/sec, starting at the beginning, to realize all the possibilities.

Does $\infty = \infty$? Not necessarily. To revisit an earlier example, there is an infinite number of odd integers, the set of odd and even integers is also infinite, and so is the set of all real numbers. All those are equal to ∞, but each set mentioned is successively larger.[8]

The subtitle implies that despite all the uncertainties and complication, I learned to love contingency. Like many, the unlimited variety of landscapes that contingency produces gives me physical, spiritual, and intellectual pleasure. But beyond that, I learned to embrace the scientific implications of contingency through the perfect landscape and related concepts.

Perfection

The perfect landscape concept tells us that we should expect idiosyncrasy and individuality in nature. Things will typically turn out to be more complicated than they first appear, and not just with respect to the details and finer points. For example, global warming is having profound effects on permanently frozen (or formerly permanently frozen) ground in Arctic and subarctic tundra biomes, generally causing permafrost thawing and degradation. This is important not only for local and regional landscape change but also because the thawing releases methane, a powerful greenhouse gas, thus exacerbating warming. Rapid Arctic warming is indeed resulting in widespread permafrost degradation. But this generalization cannot be uniformly applied to specific locations, as illustrated by the work of a research group from Tyumen, Russia, and George Washington University. They showed, using forty-three years of data from multiple sites in Siberia, that local vegetation, ice content, and soil moisture conditions can offset permafrost degradation.[9]

This calls for local, case-by-case approaches (informed and constrained by general principles and laws). Ian Fuller and Gary Brierley and colleagues illustrate this point in their study of several rivers in the east coast region of New Zealand's North Island, where (geologically) recently uplifted, highly erodible rocks, and high-intensity storms generate exceptionally high erosion and sedimentation rates. Human impacts are recent (~650 years since Māori arrival, 250 years of colonial impacts), but human disturbance has profoundly altered river systems across the region. Their study documented watershed-by-watershed variability in anthropogenic impacts in the form of geomorphic river stories for five catchments. Despite the geographical proximity of the watersheds and some common environmental controls, they show how targeted, fit-for-purpose (that is, tailored for each river), process-based rehabilitation programs are required to facilitate recovery—not quite one size fits none, but one size fits one. Though severe human-driven landscape transformation has been ubiquitous in the region, the river stories for the five study watersheds highlight profound variability in river diversity, process regimes (sediment sources, connectivity relations), and evolutionary trails.[10]

The law-place-history triad, the landscape evolution space framework, and the axioms from the preceding chapter are useful ways to understand perfect landscapes and guide responses to environmental change. Some other useful guides for this kind of place- and history-sensitive approach are Wes Jackson's *Consulting the Genius of Place*, with an agricultural focus, Kirstie Fryirs and Gary Brierley's methods for river management, the riverine ecosystem synthesis of James Thorp, Martin Thoms, and Michael DeLong, and Ruth DeFries's *What Would Nature Do?*[11]

LESSONS FOR THE ANTHROPOCENE

Landscape evolution is about environmental change—fast, slow, ancient, modern. But from the dawn of the Anthropocene, and accelerating, Earth systems are being subjected to the larger and faster changes and disturbances than ever before (and not just climate), and landscape change is accelerating accordingly. The *Anthropocene* is widely, though hardly universally, regarded as the new, current epoch of Earth history and geologic time where humans and our activities are a profound, and often dominant, factor on the planet. Despite debates and uncertainty about its status as a geological epoch, humans are unquestionably having significant impacts at all scales on landscapes and on the global Earth system.[12] What can what we've learned of nature's mysterious ways teach us about how to comprehend and manage the snowballing changes?

One thing we've stressed repeatedly is the need to consider geographical and historical contingency as well as general laws to understand how landscapes and ESS function. As we cannot complete the task with laws alone, we look for *lessons*, and we'll start with some simple ones—and by simple, I mean straightforward, not necessarily *easy* to apply.

If It Did Happen, It Can Happen

Environmental history, the geological record, and other paleoenvironmental records are valuable guides to possibilities for future environmental change. Evidence of past changes—with suitable attention to context, of course—is pretty good proof that something *can* happen again. It is on us to learn all we can about what enabled and triggered historical changes so that we can possibly forestall (or in some cases encourage) them in the future or prepare for them.

We have strong evidence, for instance, that the global thermohaline circulation has collapsed in the past and for the dramatic climate, ecological, and other changes that resulted. Given that it has happened before, under circumstances that appear repeatable, it could happen again. Likewise, roughly 14 ka, there occurred a large, rapid global sea level rise event (16 to 25 m in half a millennium or less), known to geologists as Global Meltwater Pulse 1A (mentioned above), with catastrophic (in the sense of being overwhelming) impacts on coastal and near-coastal areas.

But these events also indicate the importance of better understanding these events. For the global meltwater event, for instance, there exists uncertainty about the relative contributions of meltwater from various ice sheets. Forecasting a similar event may depend on sorting this out.

In other cases, the lessons are a bit more general. We know that various combinations of drought, overgrazing, and other land-use practices can lead to semi-arid land degradation. We know that some irrigation practices and vegetation change can trigger soil salinization. We know of circumstances where deforestation has, and could again, cause slope destabilization. We know that changes in biogeographic distributions due to climate change or human impacts (e.g., range expansion of bark beetles) can destroy or transform forests.

If It Hasn't Happened Before, It Might Happen Anyway

As outlined in the previous chapter, Earth, its biosphere, and landscapes can create novel entities. And Earth has always been, and always will be, subject to changing boundary conditions and inputs triggered both externally and internally. Therefore, while things that have happened before are a good guide to what could happen, they are not a complete guide. We have also learned that landscape evolution and environmental change often follow unpredictable paths (and that there are often many possible paths). On top of this, the Anthropocene presents novel boundary conditions, disturbances, and contexts for landscape changes and our responses to them. Tropical cyclones, for instance, are a major disturbance agent on much of the planet (mainly but not exclusively coastal areas). The rapidly warming oceans make a fine recipe for brewing up tropical cyclones that can blow up into typhoons and hurricanes, and there exists considerable evidence that storm climatology and cyclone traits are venturing into previously unknown territory. Things that have never happened before, at least during *Homo sapiens* time on the planet, will occur.

This indicates an important role for simulation modeling to explore possible future environmental responses outside the range of historical conditions and for *counterfactual analysis*. Counterfactual approaches are a type of thought experiment used in studies of the effects of (medical, psychological, policy, economic, etc.) interventions, essentially asking what would have happened without the intervention, to evaluate its effectiveness. Historians (and science and historical fiction writers) use counterfactual studies to explore the impact of specific events, such as what would have happened if a different side won a war or if a specific migration had never occurred or a given technology never introduced. As far as I know, counterfactual studies have not been applied in the landscape sciences, but perhaps they should be!

If It Can't Happen, It Won't Happen

Okay, as stated this is self-evident. But there are applicable general laws that exclude some possibilities, and conditions or resources necessary for some changes may or may not be present. More generally, environmental changes depend on a particular combination of laws, place, and history factors. Where laws forbid it, or the required place and history factors are absent, it can't happen. This is important as we seek to whittle down possible changes, developmental pathways, and outcomes.

Going back to Chapter 4, we identified eight groups from among the population of all imaginable evolutionary trajectories (or trails). Impossible ones require conditions that do not exist or are forbidden by general laws, and currently impossible pathways were possible previously but are now foreclosed by their dependence on factors that no longer exist (e.g., different atmospheric compositions, extinct biota, pre-human). These comprise the "if it can't happen it won't happen."

Improbable and inhibited trajectories may remain so or may become more likely as boundary conditions and disturbance regimes change. The category of transient pathways is dependent on temporal scale, and many of these may be quite relevant to contemporary environmental change. The recurring and selected paths are ones we should be ready for—though again, those categories may be fluid as conditions change.

Everything Is Connected to Everything Else

The key lesson of the First Law is that you can't change (only) part of a system. There will be repercussions throughout the system. This involves chain reactions, but not just one chain—in many cases, a complex web of interacting chains. Hyperconnected ESS are complex and difficult to predict, and we must bear this in mind, allowing for multiple possible trails of change, outcomes at a given point in time, and reverberating and unstable dynamics.

One of the reasons why high connectedness is so common is structural redundancy, where more than one path exists for transferring energy, resources, and information, and multiple components may be able to accomplish the same task. In ecological systems, structural redundancy refers to the extent to which more than one species (or taxonomic group) can perform a given function or play a given role in the system. Microbial ecosystems, for instance, tend to have high structural redundancy at the species level, as there usually exist multiple bacteria or other microbes that can, say, break down specific forms of organic matter, reduce iron, precipitate calcium, or what have you. Systems with a single keystone species have low redundancy, at least with respect to whatever the keystone organism does (if something else could perform the same function, then it would not be a keystone). Redundancy tends to be inversely correlated to the degree of biotic specialization and directly related to ecosystem resilience.

Structural redundancy has not been as well studied in other aspects of the landscape sciences. In some cases, it may not be very fruitful to think about redundancy. For example, aeolian sediment transport is accomplished only by wind, and fluvial erosion only by flowing water, and the disappearance of those processes (or their initiation where once absent) would be part of a larger story that is not enriched by thinking about redundancy. One could, of course, think about, say, *sediment transport* and identify multiple processes that can accomplish that—mass wasting, flowing water, waves, wind, and ice. But while the transitions among sediment transport process dominance and the possible combinations thereof are interesting and important, structural redundancy is not really something we can manipulate.

Weathering is a possible exception. Carbonate dissolution in karst, for example, has high biogeomorphic structural redundancy. Hydrocarbonate dissolution requires a CO_2 subsidy from soil and ground-dwelling biota. But any respiring plant, microbe, or animal can accomplish that. While the amount and rate of CO_2 subsidy matters, it makes no difference which organisms provide it. The same goes for bacteria that process iron (oxidation or reduction), as multiple species can perform that function (biogeomorphic equivalents or functional groups).

Sticking with karst, in central Kentucky an important process is woody root penetration into limestone joints and fractures, where a combination of microbial activity in the rhizosphere, formation of organic acids, funneling of water into the rock, and pressure exerted by root growth helps break down the rock. Only two species (in that region) seem to do this ubiquitously—*Quercus muehlenbergii* (chinquapin oak) and *Platanus occidentalis* (American sycamore). This process therefore seems to have limited structural redundancy, as root penetration of rock by other plants is less frequently observed and their root growth within the rock is less extensive.[13]

THE (IN)DELICATE (IM)BALANCE OF NATURE

The overwhelming weight of evidence is that steady-state equilibria and other balance-of-nature conditions are only one of many states that landscapes may be in, that they are rarely normative in the sense of the way things should be (and we could debate indefinitely the whole concept of "should be"), and that they are often rare and transient. This does not mean steady state and related notions are not useful reference conditions or model assumptions—they often are! But they are often misleading ideas with respect to the way ESS work. Beyond all this, descriptions often refer to a *delicate* balance of nature, implying at least implicit understanding that those conditions, where and when they exist, are vulnerable and unstable.

Thus, the mysterious ways of nature tell us that "unbalanced" conditions are not always bad and are in fact unavoidable and that our time and resources are better spent understanding and managing transitions rather than attempting to achieve and maintain elusive balances. We should also recall that stable conditions are not always good for humans and ecosystems—a severely gullied farm field, a hillside strangled with kudzu, and a dead lake choked with toxins are all pretty darn stable. As we saw in Chapter 7, instability can be a good thing, allowing for quick adjustments and state changes.

As we observe and respond to environmental changes, we cannot expect steady-state equilibrium, even after landscapes have had some time to adjust. Any disappointment or dissatisfaction that we might have should *not* be based on maintenance of a "delicate balance" or progress toward any single normative state (climax, equilibrium, etc.). We can, of course, be (dis)satisfied for other reasons.

Box 10.1

GUIDELINES FOR RESPONDING TO CLIMATE CHANGE (SEE TEXT FOR DETAILS.)

1. Do something!
2. Do something *now*.
3. Acknowledge there is *no normal* anymore.
4. *Climate changes everything*: recognize that climate change impacts everything we study, manage, plan, or design for.
5. Focus on understanding and learning lessons from "atypical" hotspots and cold spots of change.
6. Think in terms of *trails* rather than *trajectories*.
7. Embrace complexity. You can't escape it, and it must be dealt with.
8. Do *not* use uncertainty as an excuse for inaction.
9. Guide strategies with the RAD (resist-adapt-direct) framework.
10. Incorporate contingency. There are no one-place-fits-all solutions, so pay careful attention to geographically and historically contingent factors. Use (or try) fit-for-purpose solutions tailored for specific places and regions.
11. Be flexible and adaptive. In a complex, highly interconnected, constantly changing environment,

> attempted solutions may not work—or if they do, they are unlikely to work indefinitely. There are no permanent solutions.
>
> 12. Failure *is* an option. Failures are inevitable in any difficult endeavor—even more so when dealing with inherently complex systems in rapidly changing conditions with no precedents. Learn from these failures; do not let them be used as excuses for subsequent inaction.

NOTES

1. The implications of this for past-is-the-key-to-the-present and present-is-the-key-to-the-past reasoning, at the philosophical level, I will leave to others.
2. I owe this metaphor, and much more, to Robert H. (Bob) Giles (1933–2022), who taught my general systems ecology course at Virginia Tech.
3. Not to be construed as a critique of computer climate models or other environmental models, which are vital parts of the research toolbox.
4. A 2022 special issue of the journal BioScience is the best entrée to RAD. Some key articles include G. W. Schuurman et al., "Resist-Accept-Direct as a Path to a New Management Paradigm," *BioScience* 72 (2022): 16–29; A. L. Lynch et al., "RAD Adaptive Management for Transforming Ecosystems," *BioScience* 72 (2022), 45–56; S. D. Crausbay et al., "Resource Management Decisions in an Era of Ecological Transformation," *BioScience* 72 (2022): 71–90.
5. J. W. Williams, "RAD: A Paradigm, Shifting," *BioScience* 72 (2022): 13–15.
6. More detail is given in Crausbay et al., 2022 (endnote 4).
7. See Schuurman et al., 2022 (endnote 4).
8. Mathematical concepts such as *cardinality* exist to address this, but we won't get into that here.
9. G. E. Oblogov et al., "Localized Vegetation, Soil Moisture, and Ice Content Offset Permafrost Degradation under Climate Warming," *Geosciences* 13 (2023): 129.
10. I. C. Fuller et al, "Managing at Source and at Scale: The Use of Geomorphic River Stories to Support Rehabilitation of Anthropocene Riverscapes in the East Coast Region of Aotearoa New Zealand," *Frontiers in Environmental Science* (2023): https://doi.org/10.3389/fenvs.2023.1162099. This work particularly resonated with me, as I had found similar phenomena for much different river systems in southeast Texas, where rivers that were

geographically adjacent, with similar environmental frameworks, nonetheless showed quite different dynamics of Holocene change. For an example dealing with grassland ecosystems, see G. J. Brierley et al., "Development of Place-Based Catenal Models for Grassland Ecosystems of the Upper Yellow River, Western China," *Catena* 213 (2022): 106,193.

11. W. Jackson, *Consulting the Genius of Place. An Ecological Approach to a New Agriculture* (Penguin-Random House, 2010); G. J. Brierley and K. Fryirs, *Geomorphology and River Management* (John Wiley, 2006); K. Fryirs and G. J. Brierley, *Geomorphic Analysis of River Systems* (John Wiley, 2012); J. Thorp, M. Thoms, and M. DeLong, *The Riverine Ecosystem Synthesis. Towards Conceptual Cohesiveness in River Science* (Academic Press, 2010); R. DeFries, *What Would Nature Do? A Guide for Our Uncertain Times* (Columbia University Press, 2021). Also see K. R. Cox, *Geography Indivisible. How and Why Configuration Matters* (Routledge, 2023). Even if you are unconvinced by or unconcerned with Cox's arguments about unity within the discipline of geography, his points about configuration—spatial arrangement, context, and contingency—are valid.

12. The Anthropocene is an informal, rather than formally accepted, geological epoch, with some dispute about when it might have started and what it might mean. General scientific agreement exists, however, that humans are now having truly global and significant impacts. See B. D. Smith and M. A. Zeder, "The Onset of the Anthropocene," *Anthropocene* 4 (2013): 8–13; J. Zalasiewicz et al., "The Anthropocene: Comparing Its Meaning in Geology (Chronostratigraphy) with Conceptual Approaches Arising in Other Disciplines," Earth's *Future* 9 (2021): https://doi.org/10.1029/2020EF001896.

13. J. D. Phillips, "Biogeomorphology and Contingent Ecosystem Engineering in Karst Landscapes," *Progress in Physical Geography* 40 (2016): 503–526.

The Stories We Could Tell

"I t's all a question of story. We are in trouble just now because we do not have a good story. We are in between stories. The old story, the account of how the world came to be and how we fit into it, is no longer effective. Yet we have not learned the new story."[1]

The quote is from Thomas Berry (1914–2009), a cultural historian and scholar of world religions who had much to say on human relationships to Earth, and in fact called himself a "geologian." I never met the Catholic priest Berry, but I was introduced to his ideas by his brother Jim, a then-retired Air Force officer (a decorated veteran of three wars turned peace activist). James F. Berry (1917–1997) and I were both involved in environmental activism in North Carolina in the 1980s and 1990s, and Jim founded and ran something called the Center for Reflection on the Second Law (CFRSL, referring to the Second Law of Thermodynamics). I saved Jim's monthly newsletters from that era, but they somehow got lost in my multiple moves since then.

I normally shy away from the philosophical side of things—not because I think it is not important or valid but because I have trouble understanding much of it and lack patience with the semantic arguments that often seem to ignore what is (to me) the interesting stuff.[2] But the Berry brothers were well connected to real-world ground truth, and the CFRSL's motto was "we must deal with the Earth on the Earth's terms." I took that as not only a guiding principle for human-environment relations but adapted it as the un-official motto of my own research career. Reductionist, abstracted learning from the laboratory and from numerical models is impor-tant (I have benefitted greatly from them and even generated a tiny bit), but ultimately landscapes must be dealt with on their own terms in the noisy, (wonderfully) dirty, and complicated real world.

The Berrys also convinced me (or reinforced my emerging notions; hard to tell looking back over the decades) of the im-portance of stories. Communicating scientific results (to other scientists and the larger public) is an exercise in storytelling. But equally, and perhaps ultimately more, important are the larger, more encompassing stories we fit our knowledge, observations, and attitudes into. That's what Thomas was referring to in the quote at the top.

History, wrote journalist and author Tony Horwitz (1958–2019), is an arbitrary collection of facts and observations. Myths are created and perpetuated. To expand a bit in the context of his-torical Earth and environmental sciences, *history* is an arbitrary collection of facts and observations filtered by aspects of historical preservation and limitations of perception and interpretation. *Historical narratives* are created, negotiated, and perpetuated. Histori-cal narratives—explanations, chronologies, historical descriptions, chronicles, and, yes, myths—are forms of stories. The key point is that while historical (and more generally, field evidence-based) sci-ence is grounded in facts and data, however censored and variably perceived, the reporting and dissemination thereof is in the form of created, negotiated, and perpetuated stories.

Recognizing that, a group of scholars from European Union agencies addressed the need for and pathways to Earth-centric narrative stories that embrace Earth-society connections. They see narratives as "powerful means for people to associate themselves with anthropogenic global change. The starting point of our reflec-tion is that people's awareness is related to their story-telling prac-tices to communicate insights and to induce or inspire behavior." Traditional narratives were Earth-centric and served to reinforce behavior that sustained human-Earth system interactions. Though generally faith-based, the traditional thinking was functional "because it had encoded practical experiences into a solid frame of reference values." The EU group argues that modern societies centered on science and technology need to incorporate other narratives, as was the case in traditional societies. The new stories should hold relevance for mundane matters, address value creation, and reflect urban lifestyles. The mundane matters refer to aspects of Earth systems that people regularly encounter. As an example, they offer weather forecasters, who combine information of meteorological and climatological phenomena and their impacts on economic and social activities. Value creation recognizes that the application of science allows modern societies to function and provides reliable knowledge about Earth systems and their life-supporting functions. Most members of modern societies are buffered from non-human nature by engineered environments, and their lives are dominated by social interactions. The urban lifestyle criterion holds that "any Earth-centric narrative must link to these familiar society-centric narratives" to be fully understood.[3,4]

Earth-centric creation stories contrast with Western traditions inherited from Greek and Hebrew sources that are strongly committed to an anthropocentric cosmology. In contrast to the anthropocentric ideology, science—from the Big Bang theory to Darwinian evolution to the ideas espoused in this book—tells a story of continual change and creation of which humans are a miniscule (although at this time and place critical) part. Science also shows that humans are inextricably linked to—not apart from or independent of—the rest of the universe (or at least the earthly part of it).

Note that traditional creation stories need not be at odds with modern science. The Judeo-Christian traditions and story, for example, are not rejected by the Berrys—Thomas was a priest, after all. But they also accept and value science, reflecting Arthur Phillips's (1934–1989, my father) view that science deals with the physical *how* of nature, while religion deals with the metaphysical *why*. Thomas Berry writes about scientists: "Their taste for the real is what gives to their work its admirable quality. Their wish is to experience the real in its tangible, opaque, material aspect and to respond by establishing an interaction with the world that will advance the total process. If the demand for objectivity and the quantitative aspect of the real has led scientists to neglect subjectivity and the qualitative aspect of the real, this has been until now a condition for fulfilling their historical task." But now science, Earth, and society calls on us to redress those neglects—as an addition and companion to, not as a replacement for, objective and quantitative approaches.[5]

The "new story" proposed by the Berrys includes metaphysical aspects independent of science, but is nonetheless consistent with science otherwise, and is based on *creativity* (as in production of variety) and *emergence*, both of which are key themes in this book. Their new story is teleological, but as I've emphasized repeatedly, teleological goals can almost always be replicated in the form of emergence and selection. So while the new story *can* be teleological, it is not *necessarily* so. Russian geneticist and evolutionary biologist Theodosius Dobzhansky (1900–1975) is quoted approvingly on his statement that emergence is neither determined nor random, but creative. "The primordial intention of the universe is to produce variety in all things," Thomas wrote. If you replace "intention" with *tendency*—and noting that the "universe" of concern in this book is planet Earth—his view is fully consistent with mine.

What stories should we, and could we, tell about Earth surface systems and landscapes? Stories of continuous and ongoing transformative, recursive, emergent evolution; and tales of everything—including humans—connected to everything else, told with an emphasis on an integrated approach to landscape evolution. Stories of complex interactions and coevolution; of historical and geographical and scale contingency. Our narratives will include supra-organic notions and system-level evolution; they will incorporate multiple possibilities and pathways and efficiency selection. We will tell stories of adaptation and creativity. The settings will be perfect, uniquely determined by combinations of laws, place, and history.

We will tell of mysterious ways, which turn out to be not so mysterious after all.

NOTES

1. T. Berry, *The Dream of the Earth* (San Francisco: Sierra Club Books, 1988), 123. The chapter titled "The New Story," from which these quotes are taken, was first published in 1978 in volume 1 of the *Teilhard Journal*. The *Dream of the Earth* was reissued in 2015 by Counterpoint Press.
2. In this I am like many, if not most geoscientists, who if not actively resisting classification into one philosophical "-ism" or another address the issue with resounding indifference. Accordingly, Victor Baker (University of Arizona), one of the few geoscientists who has delved deeply into philosophical issues, indicates that most of us, if we are to be pigeonholed at all, are pragmatists—that is, our philosophical stances are often subconscious and focused on whatever works. In the final chapter of his 1912 book *The Problems of Philosophy*, Bertrand Russell writes that many "under the influence of science or of practical affairs, are inclined to doubt whether philosophy is anything better than innocent but useless trifling, hair-splitting distinctions, and controversies on matters concerning which knowledge is impossible." Russell then goes on to give a spirited defense of philosophical contemplation and notes that when philosophical studies reach a point where knowledge *is* possible, the subject is taken out of philosophy and placed in the sciences. I note that Earth and environmental sciences, part of the natural sciences, grew from natural history, which grew from natural *philosophy*.
3. M. Bohle, A. Sibilla, and R. C. Graells, "A Concept of Society-Earth-Centric Narratives." *Annals of Geophysics* 60 (2017): Fast Track 7, https://doi.org/ 10.4401/ag-7358. See also M. Bohle, "Ideal-Type Narratives for Engineering a Human Niche," *Geosciences* 7 (2017): 18. https://doi.org/10.3390/ geosciences7010018.
4. T. Berry also draws contrasts between the modern secular-scientific creation story and traditional narratives: "A new creation story had evolved in the secular scientific community, equivalent in our times to the creation stories of antiquity. The creation story differs from the traditional Eurasian creation stories much more than those traditional stories differ from one another" (Berry, 1988, endnote 1, 128).
5. Berry, 1988, endnote 1, 133.

For the benefit of digital users, indexed terms that span two pages (e.g., 52–53) may, on occasion, appear on only one of those pages.